FUEL
ECONOMY
HANDBOOK

FUEL
ECONOMY
HANDBOOK

Written and compiled by
National Industrial Fuel
Efficiency Service Ltd

Edited by W. Short, B.Sc., C.Eng., F.I.E.E.,
F.Inst.E., F.I.Nuc.E., F.C.I.B.S.

Graham & Trotman Limited

About the author

Formed in 1954 the National Industrial Fuel Efficiency Service Limited — NIFES — has completed 25 successful years of energy management and engineering. The objectives set down for a progressive energy conservation and planning organisation have been achieved.

Throughout the period requests for NIFES services have progressed to the stage when 200 engineers and supporting staff are engaged in work on behalf of clients. Planning and Design, complex energy utilisation studies and the continuing examination of energy strategy so vital to Industry, Commerce and Public Authorities are all comprehensively carried out by NIFES.

Reproduced from copy supplied
printed and bound in Great Britain
by Billing and Sons Limited
Guildford, London, Oxford, Worcester

ISBN 0 86010 130 4

First published in 1974
Second edition published in 1979 by
Graham & Trotman Limited
Bond Street House
14 Clifford Street
London W1X 1RD
Reprinted in 1981

© Graham & Trotman, 1979

Contents

Establish load pattern – maximum, minimum, load pattern
hour by hour.
Degree of standby capacity.
Efficiencies of various boiler types.
Typical dimensions needed to accommodate.
Typical capital and running costs (capital cost small
compared to annual fuel bills in most cases).
Combustion problems of various fuels.

Boiler auxiliaries.
Effect of short-term peaks in demand.
Chimney problems.
Automatic controls – what is worth fitting.

Possible methods – advantages and disadvantages.
Margins for severe weather.
Typical capital costs of various systems.
Methods of automatic control – savings possible.
Problems of solar radiation.
Insulation, double glazing, etc. – costs and value.

Properties – problems of storage and distribution.
Evaluation of cost effects and market situations.
Dual-fuel economies.
Residues from fuels: combustion products harmful effects,
disposal.
Use of wastes as fuels.

Foreword

Everyone now recognises that fuels and the energy they contain are amongst the most important factors in the economy and it is accepted that it is only sensible to use them in sound and efficient ways. Since the Second World War there has been a number of fuel 'crises'; a build-up of supplies just after this War, then the return of fuel oils to the market, followed in the 1958-62 period by a steady fall in the delivered prices of fuel oils. In turn this caused the contraction of the coal industry, closing of many pits then considered to be uneconomic and ever-increasing imports of crude oil for processing in refineries.

The ever-increasing demand for energy all over the world has continued without interruption for nearly thirty years since 1945, with periodic warnings from conservationists that an energy 'gap' would hit the world at some indefinite period towards the 1980-90 period. However, such vague warnings carried little weight with the majority of users when abundant supplies of cheap fuel oils continued to be available. The discoveries of natural gas in the early 1960's seemed only to herald a further period of abundance; the use of coal in gas works had already dropped to very small figures, and now even the light fuel oils and other fractions used in cracking plants which had displaced much of the original coal-based plants were released back into the market once natural gas became available. Much of this early gas was sold to relatively large users in the industrial market and within only a year or two the average price of gas to the industrial user fell by around 40% while consumption in industry increased manyfold. From being a relatively expensive fuel which could only be justified for special applications where convenience and production advantages outweighed fuel cost, gas suddenly became a serious competitor to oil and coal.

However, after what may well come to be regarded as the happy, carefree 1960's, the storm-clouds gathered. Many of the oil-producing countries began to demand greater participation in the exploitation of their oil reserves and a greater share in revenues. Almost simultaneously the great

v

DIAGRAM 1
AVERAGE COST OF FUELS TO BRITISH INDUSTRY

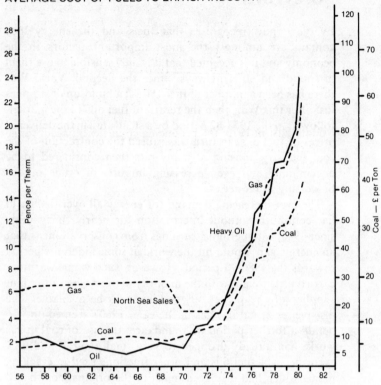

Electricity cannot be shown on this diagram as its cost in terms of its heat equivalent is completely off the scale.

Typical figures are:—
1974   26p per therm
1975   47p per therm
1976   52p per therm
1977   58p per therm
1978   65p per therm
1979   70p per therm

This diagram shows that prices of industrial fuels remained relatively constant from 1956-1970, only gas being affected at the end of this period by bulk sales of cheap natural gas. However, from 1970 all fuels have increased manyfold, oil due to the OPEC price rises, with gas and coal following due to market and inflation pressures.

vi

energy complex of the U.S.A. ceased to be self-sufficient on American sources of energy and entered the competition to secure oil from the Middle East countries. The first faint calls for energy conservation began to emerge at Government level in the U.S.A. All these pressures caused the price of fuel oils to creep slowly upwards between 1970-73 but without noticeable alarm, since inflation was causing price rises in other raw materials, commodities, manufactured goods and wage levels.

Unfortunately, under the terrible influence of War in the Middle East in 1973, the producing countries adjacent to the conflict first reduced supplies and very shortly afterwards forced crude oil prices to more than double in a matter of a few weeks. Although the reduction of supplies did not last long, the higher prices seem permanent. The diagram (No. 1) illustrates how prices of coal and oil, after remaining fairly steady for fifteen years, have risen rapidly, oil due to the producer countries and coal due to wage increases and regular attempts to reduce operating losses.

For the first time for fifteen years the average industrial price of oil per delivered therm is higher than that for coal, and both are approaching what was only a few years ago considered to be a premium price for town gas.

We now face a period when the same old mistakes can be repeated. It may well be that the North Sea finds, particularly of oil, may be exploited as rapidly as possible, using the twin excuses of the quickest possible return on the very large investment and the need to be as self-sufficient as possible on oil supplies. At the same time, and more immediately, efforts are being made to reverse the trend of coal production and stabilise or even increase annual tonnage available. If we are not careful, the development of the North Sea could be too rapid – the annual tonnage could 'squeeze' coal tonnage again and cause another depression in the mining industry. Unfortunately this would be only for a few years; the peak output of oil from the North Sea could be reached in only a few years, hold at this peak very briefly and then start a decline, causing a further energy gap to develop. Although we expect nuclear energy to develop as a gap-filler, there is a great deal of development and another very high capital invest-

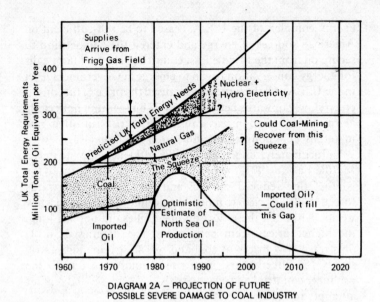

DIAGRAM 2A — PROJECTION OF FUTURE
POSSIBLE SEVERE DAMAGE TO COAL INDUSTRY

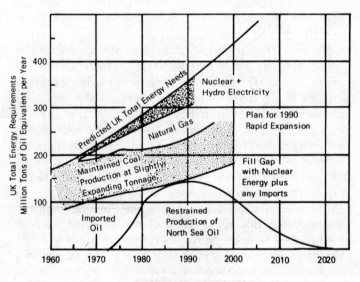

DIAGRAM 2B — HOW TO USE
OUR OWN COAL, GAS AND OIL
TO BEST ADVANTAGE

ment programme needed to produce a really significant segment of Britain's energy 'cake'. Possibly, as this is being written for the second edition in 1979, it is already too late to hope that the production of oil will be controlled to allow the coal industry to build up an accelerating programme of production for the 1980's. Apart from the commercial developers wishing to obtain the quickest return on capital we now see in 1979 an actual shortage of Middle East crude oil due to the political problems mainly in Iran. Although prophets have forecast such a shortage they were looking 30 years or so ahead. Diagram 2A shows what seemed likely before 1979, and such a situation could have pinched-out the coal industry in 1981. However, and more by market forces than long-term planning, Britain started to sell part of North Sea production into Europe, so 'flattening' the peak shown on the diagram.

Really exciting possibilities now exist for coal with the discoveries of vast reserves in the Selby and Vale of Belvoir areas providing bold capital expenditure policies are followed and wage inflation/labour availability do not hold back the build-up of output. Diagram 2B shows the fleeting opportunity that could be grasped, with a steady future of increasing coal production, expanding rapidly by 1990. Although work is going on at a large pilot scale level, cost projections of the various wave-power systems, the most likely of the 'alternative energy systems', show that electric power fed to the national grid would work out as 5 to 10 times that of oil-fired or nuclear power stations. The 1979 problems at the Harrisburg nuclear power plant in the U.S.A. may well slow down nuclear power development as environmental and political factions argue about this 'disaster, catastrophe, etc . . .' (In which saner voices point out no-one was killed or even slightly-injured and the radiation hazards almost entirely contained.)

So we have to buy time by saving fuel, rather simpler and shorter words than energy conservation. If existing supplies can be stretched over a longer period by conservation, the time is being bought for further work and research on alternative energy systems in the hope that running costs might be reduced, on nuclear power safety, nuclear waste disposal, and the only long-term hope, fusion reaction. If we do nothing to save energy, then there will not be enough to go round and commercial, political and even military actions will force prices far higher than we dream.

Accountants and governments must re-think the economic equations. Good fuel-saving ideas must no longer be rejected because the pay-back period (saving in annual fuel bills compared to capital, etc. costs) is considered in comparison with other investment options. Instead they must be reconsidered on likely fuel prices affected by inflation and politics over the next few years. No-one would have thought that the price of fuel oil would have been multiplied by 12, or the price of coal by 7, over the ten years from the start of 1970. We are perilously near to rationing, imposed either by suppliers or Governments, so that loss of profit due to reduced production must be seriously considered. Such losses can be very high as many firms only make profit on the last 10% or so production. The first 90% only produces sufficient income to pay costs such as the fixed costs of accommodation and machinery, wages, fuel, etc.

Surveys, tests and investigations by NIFES have repeatedly shown that average savings possible are around 25-30%; of course more in some industries, less in others. Indeed, savings above 30% are possible if pay-back periods are slightly extended beyond industry's present rather pessimistic ideas. A world-wide campaign can buy time by stretching existing known conventional fuel reserves for another 15-20 years. Also this reduces fuel price increases, a great benefit for developing countries. This extra time allows new energy sources to reach commercial scale and could eliminate the fear of an energy 'gap', quite apart from the financial advantages of cutting fuel bills.

In the sections of this Handbook which follow, wherever financial savings are shown in examples and illustrations, these are based on energy/fuel prices at their late 1979 levels.

1
Consideration of a New or Replacement Boiler Plant: Financial Considerations

Consideration of a New or Replacement Boiler Plant: Financial Considerations

INTRODUCTION

In most factories, office buildings or local authority blocks, a boiler plant is needed. Unfortunately, even tragically, capital cost is only too often the major factor in deciding what is installed. The cheapest boiler plant in terms of initial cost may not be the cheapest when the running costs are assessed over a period of years. The running costs are largely dependent on the efficiency of the plant and this is decided, for better or worse, at the instant the order is placed. Although improvements can be made after some operating experience, the cost is usually considerably higher than if they had formed part of the original scheme, and money wasted before these improvements has been lost for ever.

A boiler producing 5,000 kg/h of steam on a single-shift basis of 2,500h/yr will burn around £113,000 fuel per year (say oil at £100/tonne). This is twice the initial cost of such a plant, repeated every year, so that fuel running cost ought to be more important than installation cost. Sadly, to date the opposite has been true, particularly with local authorities.

DECIDING THE LOAD PATTERN

In selecting a boiler plant, the first thing to determine is the nature of the load. Wherever possible, this should be obtained by direct measurement although this is only possible where replacement boiler plant is being considered – even then the likelihood is that some projected extension has been responsible for new boiler plant being considered. Where boiler plant is required for a new factory the nature of the load must be built up from the theoretical loading of the projected process plant (at an agreed diversity factor), the theoretical space heating requirements plus an agreed allowance for

3

expansion and some safety factor (it is usually less expensive and less unpleasant to have oversized rather than undersized plant).

The facts should be tabulated in some form as below:

WINTER LOAD	Process			Space Heating	Total
	H.P. Steam	L.P. Steam	Hot Water		
Maximum possible all plant on maximum.					
Maximum likely sustained load.					
Minimum likely sustained load.					
Average day load.					
Average night load.					
Total per day.					

The exercise should then be repeated with Summer load only.

Where 'peaking' loads are encountered, it is useful to graph a typical load pattern to ensure the selected plant will have sufficient flexibility and inertia to meet the loads.

There is no substitute for a detailed appraisal of the site, load, labour, and all financial considerations. Where process loading is concerned it is important that steam metering be employed over a sufficiently long period to learn the truth. Maximum possible loads should be turned on and minimum combinations of plant studied. Process operatives frequently practice artificial economies when a test is being undertaken, the observer should be at pains to discover whether this is taking place.

It can be misleading to accept makers' figures for steam consumption of many items of plant. These would be given in good faith but may be changed radically by the user in respect of throughput, batch time, liquor quantity, liquor/product ratio.

Flow metering is a necessity particularly if packaged boilers are to be installed as the case for thermal storage can then be given full consideration. In exceptional circumstances, packaged boilers have been installed with great success in dye houses which produce an extremely difficult type of load. These were supported and protected by a suitable steam accumulator.

The thermal storage boiler is extremely valuable where thermal storage is required at high pressure which cannot be given by the steam accumulator or where space considerations will only allow the installation of one shell.

ONE BOILER OR SEVERAL

While one large boiler is usually cheaper than two boilers giving the same total output, if the seasonal load variation is very great there may be a case, on actual running costs, for a two (or more) boiler layout. Most boilers give peak efficiencies in the range 60% to 90% of rating, the reduction in efficiency below 60% depending largely on the particular design, the fall-off being less with the boilers having a large amount of convection heating surface. Even with these however, once the load is below say 30%-40% of rating, a 5% decrease in efficiency compared to that between 60%-90% of rating, is typical. For a plant operating on space heating load, an efficiency increase of roughly 2½% is thus possible by intelligent operation of a twin boiler plant, even with modern designs, and this increase would be much greater with older types. On a consumption of say 1,000 tons coal per season, this represents a saving of 30 tons or say £1,200. Over the likely life of the plant such a saving can easily repay any additional cost of twin boilers.

Quite apart from actual savings, some firms may feel that the 'insurance value' of two boilers justifies the extra cost. If a factory is dependent on one boiler, then any breakdown would mean complete loss of production, but with twin boilers and correctly duplicated pumps, fans and firing equipment, it is unlikely that a breakdown could shut down both boilers.

5

One of the twin boilers operating 'flat-out' until the other can be brought back into service might keep much of the factory in production. In some manufacturing trades, such as chemicals and plastics, this steam might enable batches to be completed and certain vital plant to be kept hot, when complete shutdown might result in costly wastage on spoilt material and further time loss after steam has been restored, in cleaning out and re-heating plant.

Even with space heating twin boilers enable some proportion of load to be carried when one boiler fails, and for most of the season the outdoor conditions are mild enough to enable virtually full comfort conditions to be maintained. Even in very cold weather the one boiler at full output may give conditions which staff are willing to accept for a short period, where failure of a single boiler may cause complete stoppage of work.

ASSESSING THE COST OF BOILER OUTAGE

The cost of 'Outage' includes the value of production lost due to lack of steam, the value of production spoiled through a shortage of steam or due to incorrect pressure and temperature, loss of amenity due to failure of space heating or domestic hot water, or even inability of staff to stay at work in very cold weather.

It is important to have some assessment of the value of outage so that one can determine the extent of standby capacity that should be installed.

It is really an insurance problem with the value of outage being the risk and the cost of standby plant being the premium. It would be desirable in all cases to be able to make a quantitative assessment of the value of outage – it would then be easy to check whether the premium is excessive.

In very few cases is a precise assessment possible. Let us consider two typical cases.

Example 1

A firm engaged in the manufacture of various bottled and canned foodstuffs and pickles. Say it has a labour force of 100 people (75% female), a wage bill of £5,000 per week, and normally operates a 40 hour, 5 day week.

Short-term hold-ups can be made good by overtime working. Medium term ones could involve some loss of foodstuffs. Long-term hold-ups could have an effect on market. What value can be put on 'outage' and what premium is it worthwhile paying?

(a) The cost of short-term outages could be calculated on the basis of overtime payments only. Assume all useful work stops with loss of steam. Cost of labour per hour =

$$\frac{5,000}{40} = £125$$

Cost of overtime to make good at time x 1.25 = £156 per hour.

(b) Medium-term hold ups. Let us regard this as anything over 5 hours. Then there could be loss of foodstuffs and the loss will increase as the hold-up lengthens – let us say that £300 worth is spoilt after 5 hours delay, £600 worth at one day delay and £1000 worth at 3 days outage – total value of food not processed before being spoiled.

We can assess the value thus:

(i) 5 hour delay – £300 food spoilt plus £780 overtime to make up production loss.

(ii) One day delay – £600 food spoilt plus £1,250 overtime to make up production (8 hours overtime).

(iii) Three days loss of steam – £1,000 worth of food spoilt plus £3,745 labour to make up production (overtime working for total of 24 hours).

(c) Long-term hold-ups – similarly a value can be put on the effect of such hold-ups on market, but allowing for labour being laid-off for the hold-up period and possible ability to avoid buying raw foodstuffs etc.

The size of boiler required was estimated at 4,000 kg/h at a total installed cost of £40,000. The cost of splitting the load between two boilers each at 3,000 kg/h was £60,000. The total extra annual cost of using two boilers instead of one was estimated at £4,000 per year. (Extra capital repayments and interest costs, maintenance, insurance, etc.)

7

We have got an approximate value of losses incurred by breakdown and the cost of the standby 'premium'. What are the risks of breakdown? From NIFES statistics a possible pattern can be suggested:

Chances of breakdown exceeding 1 hour but less than 5 hours – say 3 per year maximum.

Chances of 24 hour breakdown – say one every 2.5 years.

Chances of three days or over breakdown say one in 6 years.

Value of lost production in any year is, therefore:

3 at 5 hours at £1080	=	£3240
1/2.5 at 24 hours at £1850	=	£ 740
1/6 at 3 days at £4745	=	£ 790
		£4770

......

The 'premium' is, therefore, £4,000 per year for·a likely loss of £4,770 per year. On this basis there is no real case for standby plant, as even the solution of two 3,000 kg steam/h boilers would not have allowed full normal production if one had suffered breakdown. This example could of course be made more detailed by taking into account fixed costs, such as rates, rent or lease of building, administration costs, etc., which continue even if no production is taking place and these might just improve the case sufficiently to choose two boilers.

Example 2

A hotel has a load as follows:

Space heating 30 Therms per hour.

Cooking 900 kg steam per day mainly occurring 11.30 a.m.-2.00 p.m. and 6.00 p.m.-9.00 p.m.

Domestic 10 Therms per day, load frequency of
hot water 1/3 between 7.00 a.m. and 10.00 a.m.
1/3 between.
1/3 between 6.00 p.m. and midnight.

What size boiler plant, what number of boilers and what is cost of 'outage'?

The cost of outage is almost impossible to calculate. It is loss of amenity largely and it would depend on the severity of the weather to some extent. The management of the hotel may not care to value the cost of outage, but would say that they could not tolerate loss of steam since their main meals are completely dependent on this.

The likely maximum load on the plant is about 2,200 kg/h of steam in Winter and 1,000 kg/h in Summer. The minimum steam load is likely to be about 400 kg/h Winter nights and nothing Summer nights after midnight.

It would appear that two boilers, each rated at about 1,600 kg/h would be adequate. These two boilers might cost £30,000 to install, compared to £20,000 for a single 2,400 kg/h boiler, and the extra annual running costs of around £2,800 for extra capital and interest payments, maintenance, etc., could be offset by a minimum 5% fuel saving, about £1500 using light fuel oil, so the 'premium' for insurance is £1,300 per year.

If we take the same likelihood of boiler breakdown as before, we can say there is likely to be three ½ day breakdowns on each boiler per year, one full 24 hour stoppage every 2.5 years, etc. So the likelihood of having no boiler at all available on any day is very remote and can be discounted. There are only about 20 very cold days in each year when the loss of one boiler would be felt so there is a 3 to 1 chance against such an occurrence in any year. Even if it did occur, the effects could be minimised by judicious adjustment of space heating in corridors and unused rooms.

FUTURE LOAD MARGIN

Any new boiler installation may be considered with future expansion of process, or extension of space heating load due to additional buildings in mind. It may be tempting to install a plant capable of meeting this future load rather than immediate requirements, and if this future load is only a small percentage on top of existing requirements this may be quite sound. However a large extra load, which may not occur for several years (and which may never occur if policy or future business alters) may cause a considerable amount of extra

9

capital to be locked up in boiler capacity which can serve no useful purpose until the extra load occurs.

It may be better to consider a twin-boiler installation, installing only one boiler at present, either building a boiler-house large enough for the second boiler, or leaving space at the side of the boilerhouse to allow it to be extended. The size of the anticipated load may determine whether the boilers should be 50:50 or some other rating ratio. Alternatively three equal boilers may be considered, installing two immediately and allowing space for an eventual third.

Apart from considerations of efficiency and availability of some output in case of a breakdown, a multiple boiler installation does have advantages when insurance inspection and maintenance are considered. With a single boiler, all work on it which involves stopping steam or hot water generation has to be done at a weekend or holiday shut-down if load is required all year. With a multiple installation, work can be spread out during the light load period and usually carried out at ordinary wage rates instead of overtime or weekend rates.

The question of sudden peak demands on steam boilers can also affect choice, but these peaks deserve attention outside the boilerhouse, and will be discussed later in a subsequent section.

BOILER EFFICIENCY

NIFES have carried out some thousands of tests on boilers of all types, and of all ages, including many new installations. The results demonstrate two clear points. First, the average level of boiler operation in this country is lower than it should be. Second, new plants do not necessarily operate at high efficiencies unless the operator is properly trained and the plant is well maintained. It cannot be over-emphasised that good plant will not, of itself, save fuel unless it is operated and maintained by competent staff who know what they are doing. A basic amount of instrumentation is essential, although it is regrettable that at many firms instruments had been provided but were out of action or giving false readings due to lack of proper servicing. Without measurement of combustion conditions an operator tends to set burners or stokers to use too much excess air in order to give a clear chimney top, but unfortunately wastes fuel.

Consider two small steam boiler plants of identical design, one run well and the other just allowed to run, being considered just as a service to the factory. It is assumed that the load is around 80% of the design rating of one boiler, that fuel oil is used at 30p/gallon, and that each individual boiler is rated at 2,000 kg/h of steam.

ITEM	Well Run Plant	Indiffer- ently Run Plant
No. of boilers on line	1	2
Load on each steaming boiler (% M.C.R.)	80	40
Feed water temperature (°C)	72	39
Percentage of condensate returned	80	40
Flue gases at boiler outlet:		
CO content (%)	12.5	9.5
Temperature (°C)	210	218
Excess combustion air (%)	28	65
Heat Account (based on Gross C.V. of fuel)		
Dry flue gas heat loss (%)	7.8	10.7
Heat loss due to H in fuel (%)	6.8	6.9
Radiation heat loss (%)	1.2	2.5
Allowance for unmeasured heat losses (%)	1.5	1.5
	....	....
Total heat losses (%)	17.2	21.6
	....	....
Inferred gross thermal efficiency (%)	82.8	78.4
	....	....

	Btu/lb	kJ/kg	Btu/lb	kJ/kg
Total Heat of steam at 100 lbf/in²g and 0.97 dry (690 kN/m²)	1,164	2,708	1,164	2,708
Heat of steam above feed	1,035	2,408	1,095	2,543
Fuel consumed to produce steam.	lb./1,000 lb. or kg/tonne		68.3	76.2

The indifferently run plant has dropped 4.4% in efficiency; a 5.3% waste of fuel, and overall the steam is costing 11.5% more in fuel costs to produce. The annual

11

fuel bill of the well-run plant could be £40,000 on two shift working, and for the indifferent plant £44,600, or a loss of £4,600 and an unnecessary fuel oil consumption of 15,600 gallons per year.

Diagram 3 shows actual test figures for four makes of packaged steam boiler, all of similar size and price. All have efficiencies above 80% at full load, but type D would be a very poor choice for a space heating load where for most of the heating season it would be operating at part load only. Boilers B or C would be a much better choice – if the boiler load over the whole heating season averaged 45%, the difference in efficiency of 7% represents a 9% fuel saving. Boiler D would be more suitable for a process plant with a steady load fairly near to the full load rating. Makers should be asked to give guaranteed efficiency curves over a considerable load range.

The lower part of the diagram shows the excess air used for combustion at various loads, as measured by the oxygen content of the flue gases. It can be seen that the efficiency of these oil-fired boilers is largely dependent on the performance of their burners at part load conditions. Most burners are designed to give good combustion at full load with relatively small excess air usage, but some burners do not give complete mixing of air and fuel if the air velocities through the burner register are reduced in the same proportion as the fuel flow. The burner on boiler D is particularly bad, with hardly any modulation of air flow.

Diagram 4 illustrates, for a coal-fired boiler, how by tests an optimum excess air usage can be found which gives the best boiler efficiency. Below this optimum, efficiency drops due to incomplete combustion (formation of carbon monoxide and hydrogen which remain unburned in the exit gases and an increase in unburned material in the ashes and clinkers). Above this optimum more heat is carried away by the increasing amounts of excess air. It can be seen that if by indifferent adjustment the excess air figure is allowed to rise to 60% the boiler efficiency drops by 4.5%, representing a 6% loss of fuel. In fact this diagram shows test results for a particularly good coal fired packaged

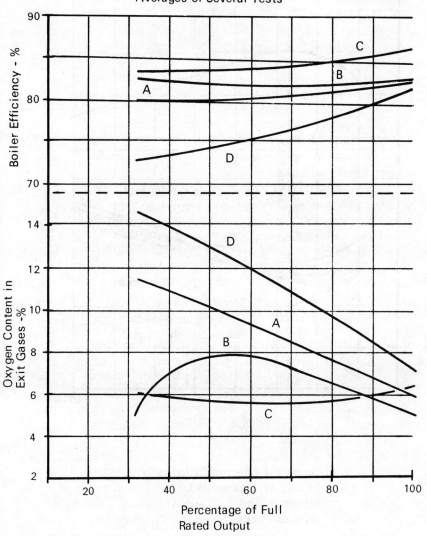

Diagram 3 Efficiencies and
Combustion Performance of
Four Packaged Steam Boilers-
Averages of Several Tests

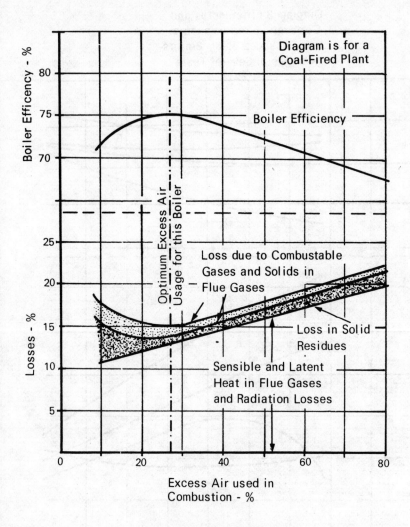

Diagram 4 Effect of Differing
Excess Air Usages on Boiler
Efficiency and the Various
Sources of Loss

DIAGRAM 4A
AVERAGE LOADS OF TESTED BOILER PLANTS

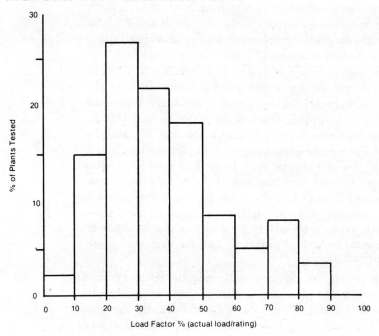

Load Factor % (actual load/rating)

SUMMARY OF COMBUSTION SURVEYS ON 100 BOILERS
(of each fuel)

	% CO_2 in Combustion Gases		
	Full Output	Half Output	Maximum Theoretical
Oil-Fired	10	8	15.5
Gas-Fired	9.5	8	12
Coal-Fired	9.0	7	18

Remember, very few boilers work at full load all the time; indeed space heating boilers may spend nearly all their working life operating at half load or less. The diagram shows the average loads (over actual working hours) for several hundred boilers tested by NIFES over the last few years. This makes the results obtained recently by NIFES on tests made on 100 boilers most interesting. On average the combustion settings were poor, particularly at half-load, and as very few boilers operate at full load continuously, the half-load settings may be a better guide to fuel wastage. Pay a little more if necessary, for burners which give good smoke-free and complete combustion at part-load conditions. If your existing burners are just incapable of being adjusted, or cannot be modified at reasonable cost, to give good combustion with minimum excess air, then it will pay to scrap them and replace.

15

boiler, and it is more usual for the optimum excess air figure to be around 40% for coal firing. With oil firing and a good modulating burner similar to types B and C in the previous diagram, typical excess air figures of 30% to 35% give optimum efficiencies.

Even after choosing a boiler with a flat efficiency curve to prevent fuel wastage at part loads and after ensuring the firing equipment is set to optimum excess usage, Diagram 5 reinforces once again what has already been stated, that the major factor in the cost of heat is the fuel cost and the capital charges (capital repayment and interest) are only a small part of the total cost. If by increasing the capital charges by 10%, the efficiency of the boiler plant can be increased so as to save 2% of the fuel used, then this is a good investment; for example, from Diagram 5 for a boiler with a usage factor of 25%, the capital charges would be increased by only £0.10/tonne but the fuel cost would drop by £0.20.

Of course this diagram really only applies to a certain range of boiler plant size; as plants become larger the total cost of steam drops due to scale effect, but not too greatly, and the drop is mainly due to a relative decrease in capital charges and labour costs. The diagram also illustrates the major influence exerted by usage factor, the number of hours per year combined with percentage of boiler rated output (e.g. a boiler operating three shifts, five days per week, on a steady load of 80% of rating would have a usage factor of 60% and at the other end of the scale a single space heating boiler for an office block where the staff work a 40 hour week may have a usage factor of only 10%). Diagrams 6 and 7 show how the fuel cost of steam can be estimated for fuel oil and coal.

ESTIMATING INSTALLED COSTS OF BOILER PLANT

Budget costs are often wanted for investment appraisal and capital allocation well in advance of any detailed design work or tender prices being available. The prices of boilers themselves are quite readily available on enquiry from

16

DIAGRAM 5
COSTS OF STEAM FROM TYPICAL BOILER PLANTS (late 1979)

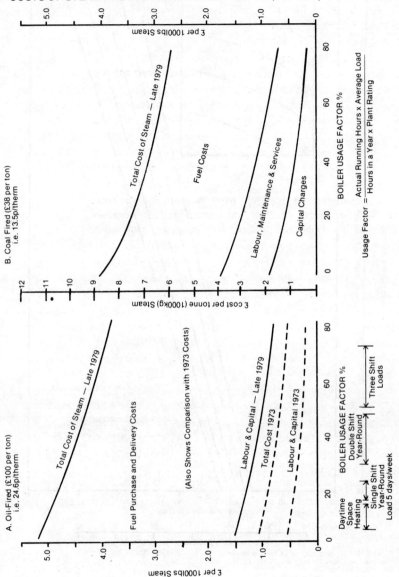

A. Oil-Fired (£100 per ton)
i.e. 24.6p/therm

Total Cost of Steam — Late 1979

Fuel Purchase and Delivery Costs

(Also Shows Comparison with 1973 Costs)

Labour & Capital — Late 1979

Total Cost 1973

Labour & Capital 1973

£ per 1000lbs Steam

BOILER USAGE FACTOR %

Daytime Space Heating

Single Shift Year-Round Load 5 days/week

Double Shift Year-Round

Three Shift Loads

B. Coal Fired (£38 per ton)
i.e. 13.5p/therm

Total Cost of Steam — Late 1979

Fuel Costs

Labour, Maintenance & Services

Capital Charges

£ cost per tonne (1000kg) Steam

£ per 1000lbs Steam

BOILER USAGE FACTOR %

Usage Factor = Actual Running Hours x Average Load / Hours in a Year x Plant Rating

DIAGRAM 6

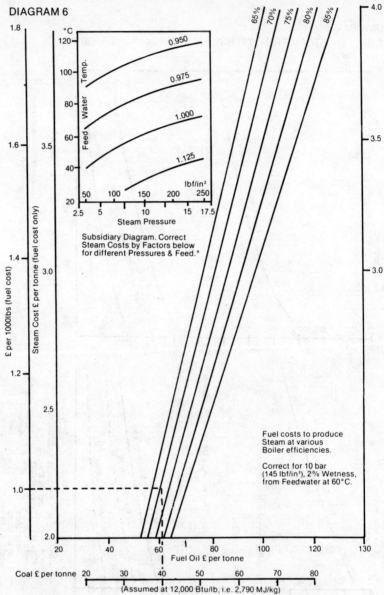

Subsidiary Diagram. Correct Steam Costs by Factors below for different Pressures & Feed.°

Fuel costs to produce Steam at various Boiler efficiencies.

Correct for 10 bar (145 lbf/in²), 2% Wetness, from Feedwater at 60°C.

£ per 1000lbs (fuel cost)

Steam Cost £ per tonne (fuel cost only)

Fuel Oil £ per tonne

Coal £ per tonne

(Assumed at 12,000 Btu/lb, i.e. 2,790 MJ/kg)

EXAMPLE: With coal at £40/ton, a Boiler at 70% efficiency, the fuel cost of Steam would be £1.00 per 1000lbs (£2.20 per tonne).

18

DIAGRAM 7

STEAM PIPEWORK FROM BOILERS

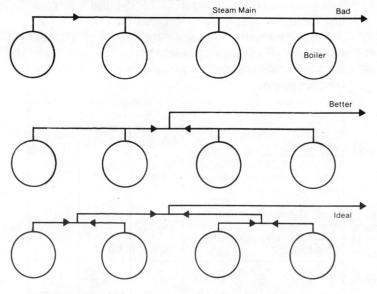

With multiple boilers connections to the steam main feeding consumers can be done badly and the boiler nearest the load will always respond faster, often causing serious feedwater and load hunting problems. It may also be difficult with bad layouts to obtain the maker's full rated output from the boiler at the far end of the common main as it may have to operate at slightly higher pressure to overcome the extra pipework resistance. If the other boilers are up to normal pressure this will cause the burners of the first boiler to modulate down on individual pressure control.

In amplification of Diagrams 5 and 6, a recent series of tests carried out by NIFES on 76 plants illustrate that the costs of steam delivered to consumers, exported from the boilerhouse and generated by the boilers can all differ on the same plant due to losses in transmission and in auxiliary usage.

Shift Pattern	Number in Sample (1978)	Total Cost of Steam Generated £/1000lbs	Total Cost of Steam Exported £/1000lbs	Total Cost of Steam at Consumers	
				£/1000lbs	p/net therm
Single	23	3.18	3.34	3.87	39.9
Intermediate	41	2.62	2.78	3.16	32.3
Three Shift	12	2.21	2.41	2.72	28.0

These follow the expected pattern with the cheapest being three shift and the most expensive being single shift, and all fit within the parameters of Figures 5 and 6.

19

manufacturers and in spite of local variations in other costs due to site problems or local planning requirements, examination of large numbers of detailed costs for installation of boiler plants suggests that the following multiples are sufficiently accurate to give first order accuracies. The following Table gives these multiples expressed as fractions of the actual cost of the 'packaged' boilers themselves.

ITEM	Multiples times Boiler Cost		
	1 Boiler	2 Boilers	3 or more Boilers
Boilers	1.00	1.00	1.00
Flue gas ducts from boilers	0.06	0.07	0.07
Fuel system (see footnote)	0.24	0.21	0.19
Pipework, pumps, valves, feed system, etc. within boilerhouse	0.46	0.41	0.39
Electrical wiring, panels, lighting, etc.	0.15	0.14	0.13
Sub-total, without civils or chimney	1.91	1.83	1.78
Civil work outside boiler-house (for chimney, fuel storage, etc.).	0.22	0.20	0.18
Chimney(s)	0.29	0.26	0.23
New boilerhouse	0.43	0.37	0.33
Total	3.05	2.66	2.52

Note:
(1) The sub-totals should be used when replacing an existing boiler plant by a similar one, and the full total is for a completely new plant, including boilerhouse and chimney.
(2) For natural gas fired plants on non-interruptable tariffs, omit the fuel system multiplier. Retain it if a dual fuel system is to be installed to burn fuel oil when supply of gas is interrupted.

The provision of adequate storage and conveyor systems for coal firing is usually more expensive than for fuel oil storage, but as the price of coal-fired boilers tends to be around 30% (large plants) to 45% (small plants) more than for oil or gas-fired plant, the Table automatically takes care of this. These percentages allow for flue gas cleaning/dust arresting plant and ash removal systems as part of the boiler price. Similarly hot water central heating boilers need less ambitious pipework, make-up arrangements, etc., than steam boilers, but as they are usually considerably cheaper the Table applies the compensation.

The Table should not be used as it stands for water-tube boilers; in general the cost of these is much greater than for shell-type boilers.

We have almost lost the generation of factory engineers and boiler operators used to every day operation of coal-fired plant; indeed many of the manufacturers have either gone out of business or have amalgamated with others into a more compact group making only oil or gas-fired boilers. Obviously the first increases in coal production can be taken up in power generation, particularly by reducing fuel-oil usage and we may yet see a few more large coal-fired power stations being built. However, as the nuclear option could become attractive to the electricity industry, further expansion of coal production can only be utilised in industry. Obviously some coal could be converted to oil or natural gas equivalent at a cost that at present seems unattractive, but industry can replace time-expired boilers by new coal-fired ones at more economic cost.

2
Consideration of a New
or Replacement
Boiler Plant:
Operational Considerations

Boiler Plant
(Continued from Section 1)
Operational Considerations

LABOUR

One subject on which the most extreme claims have been made by boiler manufacturers and fuel suppliers is the amount of labour needed to run a modern boiler plant. Much can be done by automatic equipment to control the plant and handle fuel, but experience has shown that this automation really creates a new problem, the need for a trained man to remain on call near the plant in case any failures sound alarms. Full time attendance is not necessary except in larger boilerhouses, but it is difficult to combine speedy response with other useful work. One possible arrangement is at fairly large works with a central repair and maintenance workshop near the boilerhouse. A fitter on each working shift can thus be trained to operate the boiler and can spend the remainder of his time in the workshop.

Even oil or gas-fired plant does require appreciable attention, contrary to the claims of some manufacturers. The various controls and alarms must be tested daily, various filters need cleaning, burners need checking and oil nozzles cleaned or changed, general 'housekeeping' and checking performance, fuel and chemical deliveries. All this time should be properly charged to the running of the plant. For smaller boilers, such as those generating one or two tonnes of steam per hour on a single shift basis allowance of at least one hour per day is realistic, rising to say one hour per shift for larger boilers on continuous operation.

With solid fuels labour allowances have to be increased, largely due to bunker trimming and ash handling. Even with small plants, fairly cheap screw feed conveyors feeding into large hoppers should be used to eliminate hard manual work of fuel handling; but these require level controllers to be fully automatic.

25

These have been temperamental in the past if relying on mechanical action, but more recently small radiation type level detectors have become cheap enough, and safe enough, to use. One cheaper compromise is a clock controller to run the elavators for a short period every few minutes, with the attendant resetting the running period according to actual load and levels in hoppers as necessary on his periodic visits.

Automatic combustion control can take care of the actual burning of solid fuel but even moving stokers, of the traditional coking or chain grate types only dump ashes and clinkers at the end of the grates. Some work was done in the 1950-1960 period on ash extractors but these normally only brought the residues to the fronts of the boilers to containers or to chutes to cross conveyors, usually water-sealed to cool the ashes. However for small boilers the cost of fully automatic disposal to overhead hoppers for discharge into lorries was prohibitive. Consequently the labour allowances should be double those for oil or gas-fired plants; e.g. two hours per shift on single boiler plants with solid fuels having ash contents below 10%, plus an extra hour for each extra boiler being used or for fuels of higher ash content.

PEAK LOAD PROBLEMS

As old boiler plants in process industries, usually of the Lancashire or earlier types of Economic pattern, have been replaced by modern highly rated boilers, usually 'packaged' designs, troubles have been experienced which have often been blamed on bad design. These troubles have occurred at high load periods, particularly when there has been a fairly rapid increase or 'peak' demand when embarrassing situations have occurred where the new boilers appeared unable to meet load even though it was apparently within the manufacturer's catalogue rating.

The following table compares heating surfaces and water contents of typical boilers and should assist in explaining some of the problems.

First it will be noted that the modern boilers are not cut down as regards heating surface, which is more generous and better arranged than on the older boilers being replaced. Indeed the evaporation per m ^{2} on the old Lancashire type

BOILER	Rating-t/hr (kW)	Heating Surface -m(2)	Evap oration kg/m(2)/hr.	Steam Output/ m(2) water surface	Water Cont- ent(t)	Boiler Shell Length(m)
Lancashire 30x8 6 dia.	5.05 (3167)	108	47.0	249	17.5	9.2
1938 Economic 2 pass	5.20 (3270)	178	29.0	390	10.5	4.9
1938 Economic 3 pass	5.0 3270)	182	28.5	479	8.9	3.3
1970 Packaged Economic	5.40 (3412)	190	28.4	806	8.5	3.5
1970 Packaged Economic	16.00 (10,235)	560	28.6	1190	22.2	5.5

boiler was higher, the reason being that this boiler was less efficient and had a high exit gas temperature of around 450° compared to the 230°C of modern boilers, so the average temperature differential across the metal surface was greater. Essentially this was a 'radiant' transfer boiler.

With modern boilers there has been a tendency, particularly with oil firing, to reduce the size of the furnace tube since recently-developed burners can produce a more compact flame. This has caused the maximum heat transfer through the parts of the furnace tube adjacent to the actual combustion volume to increase, some boilers operate on a peak transfer rate of over 600 kW/m^2 which is over three times the peak rate possible over the grate of a coal fired Lancashire boiler.

One other marked change is the water surface area in the boiler from which steam flows, and the Table shows that the steam production /m^2 water surface has more than trebled for packaged boilers of equivalent output. Much larger outputs are possible from single packaged shell boilers, one large boiler being equivalent to three or four old Lancahire boilers, but the steam rate / m^2 increases further to nearly five times, as is shown for the large boiler in the last line of the Table.

Lancashire boilers were very forgiving so far as feed water quality was concerned. Reasonable steam quality was obtained with boiler water containing 5000 - 15000 p.p.m. of total dissolved solids, and scale formation could be controlled to some extent by treatment with boiler compounds which

27

inhibited deposition, or modified the scale so that much fell down as sludge and with the relative lack of circulation at the bottom of the boiler could often lie there without harm. The older Economic boilers operated reasonably with 4000 p.p.m. but modern compact shell boilers usually have problems if the t.d.s. exceeds 2000 p.p.m. In most areas this means that the boiler can no longer be treated as a chemical reaction vessel, adding boiler treatment compounds in the hope of averting scale deposition. Unless a very high proportion of the steam output can be returned as pure condensate to the boiler, external water softening plant is essential. The additional capital cost should be considered as an investment to produce purer steam from the smaller water surfaces of present designs, as a fuel-saver by eliminating, or very much reducing, scale formation on heat transfer surfaces, and by reducing the amount of blowdown needed to control the t.d.s. in the boiler water.

Without such external treatment, blowdown quantities can be high, reachig 10 to 15% in 'hard' water districts, and unless steps are taken to recover heat from this blowdown the heat loss can represent 2 to 3% of the fuel used in the boilers. Where blowdown is unfortunately necessary to maintain correct t.d.s. values, this should be a continuous system, with valves capable of being set accurately so that by regular testing and re-setting of the valves a reasonably steady t.d.s. can be achieved. Continuous blowdown should always be accompanied by a heat recovery system. In its simplest form this could be a pipe coil in the feed tank (but check periodically for corrosion and leaks) and for larger plant more elaborate equipment, such as a flash vessel to give low pressure steam followed by a heat exchanger to reheat feed water, might be jusitified. If after all such devices, the blowdown water is still quite hot, consider other possible uses, such as heating process or domestic warm water.

With recent increases in costs of fuel, even existing heat recovery plant for blowdown water should be re-examined. The aim should be to discharge blowdown to drain at only a few degrees above the temperature of the incoming cold water into the factory or building.

DIAGRAM 7A

HEAT LOSS IN BOILER BLOWDOWN

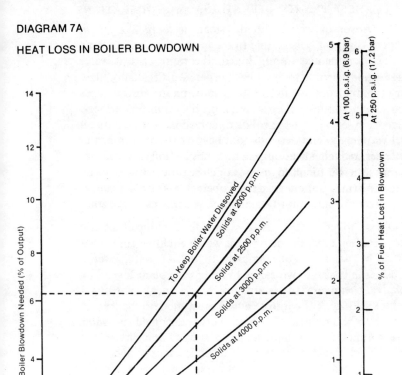

EXAMPLE: A boiler fed with water containing 150 p.p.m. of dissolved solids needs 6.3% blowdown to keep boiler water concentration to 2500 p.p.m. and blowdown contains 2.0% of fuel C.V.

This diagram illustrates the amount of energy taken out of boiler plants by the blowdown, whether intermittent or continuous, necessary to control the total dissolved solids in the boiler water below the level at which foaming or priming might occur. Previously the effort and cost to recover this heat has often been considered unnecessary, but with the escalation in fuel costs over the last few years, there is now usually a very good financial case for recovery of the maximum amount of heat from this blowdown water.

Carry-over of water with the steam is more likely with modern compact boilers, due to the lower water release area already commented upon, due to the more critical water quality and also due to oil or gas burners which can modulate very rapidly from low to full firing rate and so cause changes in water circulation patterns round the heat transfer surfaces. Such carry-over, with the solids dissolved in it, is an indirect fuel waster, as solid deposits foul heating surfaces in heater batteries, reducing heat output, and may so foul valve seats as to make them difficult to operate or close fully. Some types of steam trap may be prevented from operating efficiently and so leak excessive amounts of steam into the condensate system.

Older boiler plants used steam feed pumps, usually reciprocating type, the delivery of which could be controlled either by hand or automatically to deliver water over a considerable turn-down range to match the plant load. The relative cheapness and simplicity of electrical centrifugal pumps with on/off control caused most modern 'packaged' boilers to be supplied with these. This system held the water level within safe limits but it made no contribution whatsoever to efficient steam generation. The boiler so equipped was merely topped up at irregular intervals and there was every likelihood that the onset of a heavy steam demand would coincide with the sudden opening of the control valve causing quenching of the boiler with a large quantity of incoming feed water, often at comparatively low temperature. This is bad for multiple boilers delivering to a common steam range. There is experience of severe 'hunting' caused by such an arrangement, and the explanation is somewhat as follows:- Assume two (or more) boilers working to a common steam main and that at one instant all the individual feed pumps are stopped. As the water levels drop, due to steam generation, one level controller is bound to cut in first, and water is pumped in at full delivery rate, chilling that particular boiler, reducing its evaporation and pressure. This temporarily throws a greater evaporation load on the other boilers, and then the first pump may stop and another start, causing another load swing and a general 'hunting' effect. Also the effects of peak loads can be accentuated, since a situation can arise where all the pumps happen to be stopped as the peak comes on, and in a few seconds due to increased evaporation all will start, chilling the boilers with cool feed just as the rapid increase in evaporation is needed, and so increasing the thermal shock, which can cause 'priming' and damage to boiler tubes.

A better system would be to have feed pumps running continuously and connected to a common feed water main leading to all the boilers. Each boiler would then have a feed valve controlled by level, operating so that water is always entering the boilers in accordance with the evaporation. This would prevent the load 'hunting' mentioned above and would also reduce the thermal shock loads on sudden peaks. Suitable modulating level controllers are made by several manufacturers.

An on/off pump must deliver more water than the boiler rating, to enable the water level to be recovered when the boiler is 'flat out'. A normal minimum extra delivery is 25%, and may often be greater as manufacturers may use a single pump size for two or three ratings of boiler. Pump failure on the on/off system automatically deprives the user of that boiler, but with the system described above, using the same two pumps on a common main with modulating feed controllers, failure of one pump would still allow both boilers to run up to a load 25% greater than one boiler. If desired one standby pump could be added, which would then enable full loads to be obtained from both boilers with any two pumps in use.

With the independent on/off system with pumps 'packaged' on boilers, it is most difficult to envisage one pump acting as standby for either boiler without involved pipework, changeover valves and changeover electrical controls from each boiler level switches.

ANTICIPATORY CONTROL FOR LOAD CHANGES
Most modern boilers operate by pressure control, i.e. change in boiler steam pressure operate the control gear to increase or decrease firing rate, or in very small plants to operate an on/off firing system (e.g. underfeed stokers, oil burners).

31

It is preferable, if at all possible, to have a modulating burner or stoker in which the firing rate is varied in proportion to load (or on very small plants a high/low/off system rather than on/off). Most controlls are parallel type, i.e. the air and fuel supplies are linked to a single operating motor, and there is no feed back from one to the other. On larger boilers it may be worth investigating the admittedly more expensive controllers which have a 'series' action since these may give more consistent combustion conditions with changing wind conditions, outdoor temperatures etc. which may cause chimney draughts to alter.

Where sudden peak steam demands may occur, the pressure control system suffers from the fault that the increased load may exist for some seconds or minutes before the pressure starts to fall and the boiler firing rate increases. This time lag may cause a considerable drop in pressure before the position is restored, even though the peak load may actually be within the rating of the boilers.

In such a case it would be advisable to have an anticipatory control fitted such that a change in steam flow to the works would alter the burning rate of fuel before the boiler pressure reacted.

Such a control system is marketed by several firms, and consists in essence of a steam flow meter operating with an orifice plate. A sudden increase in steam demand would cause the oil burner controllers to increase the oil burning rate. The normal pressure controller would then override this if necessary. Such a system would give the vital few seconds warning of load change prior to boiler pressure change, reduce rate of change of heat input and so reduce thermal shock and chance of priming. We feel that in any event a master controller on the steam main should normally control all boilers and burners.

With each boiler on independent pressure control it will never be possible to set all of them to be completely in step, so that a sudden peak load may cause the boilers to react in sequence. With a master controller, all would respond simultaneously sharing the load change amongst all three boilers and preventing uneven load distribution or shock. A changeover switch would return each boiler to its own pressure control, or enable it to be shut down, if desired.

Such anticipatory controllers, whether on single boilers or controlling a range of two or more, also have advantages on sudden drop of load, detecting this some vital time before boiler pressures increase and so reducing firing rate a little earlier. Particularly with solid fuel fired plants this might eliminate safety valve losses when loads change almost instantly from full boiler output to very low figures.

Another approach to the problems of load variation and breakdown possibility with large steam boilers of the Economic type is to have a twin furnace flue boiler with independent combustion chambers and smoke tube sets. In effect there are two independent sets of heating surface contained in one boiler shell. The boiler flue outlet can be similarly divided, with a damper for each set of heating surface so that one side of the boiler can be shut down and the stoker or burner opened up for repair or maintenance without any air rushing through to cool the boiler, and the other side can steam on.

Thus a burner failure only causes half the boiler capacity to be lost, and each burner should be fitted with a fully independent control panel, again to minimise the effect of failure. Each divided flue outlet damper should have an electric interlock to its associated burner to prevent the burner operating unless the damper is open.

CHIMNEYS, CORROSION AND FATIGUE

Corrosion is important where the fuels being used contain corrosive elements such as sulphur or chlorine. This is particularly serious with plants using heavy fuel oils which can have fairly high sulphur contents up to 4%, especially from Middle East crudes, but acid gases can also be formed from light fuel oils and solid fuels. A small proportion of the sulphur is converted to sulphur trioxide which in turn can condense out in the presence of water vapour to give a strong solution of sulphuric acid. If the internal surfaces of flues, fans or chimneys are not kept above 150°C this acid deposition

33

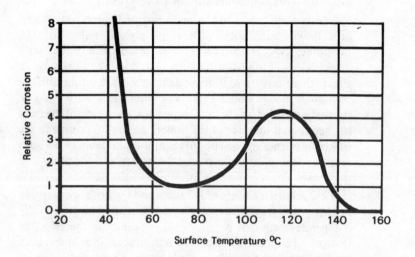

Diagram 8 — Variation of Corrosion with Surface
Temperature of Steel Flues or Chimneys

will occur and cause rapid corrosion. See Diagram No. 8. Chimneys constructed from glass fibre, with heat-resistant resins, are now available and are suitable for gas temperatures up to approximately 450 deg.F. The material is a better insulator than steel so that some reduction in condensation is achieved, but the main advantage is that no corrosion occurs, even with a film of acid condensation. Such chimneys are much lighter than steel ones. However, smut emission could still occur using heavy fuel oils and a glass fibre chimney.

The total cost of an insulated steel chimney may be over double that of an ordinary one, but the life may be increased four or five times.

Fatigue cracking due to wind oscillation has also been noticed. Diagram No. 9 is of one case tested in some detail showing that even with flue gases at nearly 500°F the internal metal temperature of the chimney was well below acid dew point for the whole height above the boilerhouse roof. Insulation by cladding with a closed air gap caused metal temperaures to rise above dew point, although only just above at the tops of the chimney. Another cause of reduced gas temperatures in chimneys is air infiltration. Where this occurs through caps in flues these can quickly be sealed, but often hidden infiltration occurs through spare boilers. It is possible to obtain gas-tight isolating dampers, and whenever several boilers share a common chimney these should be used, as otherwise surprisingly large volumes of air may be drawn through spare boilers to increase the quantity and reduce the temperature of the chimney gases. Even with a single boiler fitted with an oil burner, if this is operating on an on/off basis, excessive air quantities may be drawn through during the 'off' periods if the forced draught damper or exit damper is left open. Unfortunately, very few burner units do close these dampers effectively during these 'off' periods and both efficiency, draught conditions and corrosion troubles may be worsened. When the burner controls initiate the starting up procedure there is and essential purging period during which air is passed through the boiler without any oil flowing to the burner. It can only be said that on most boilers examined this

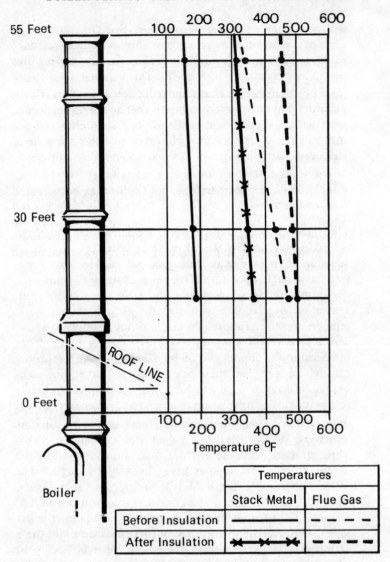

Diagram 9 — Temperature Distribution in Metal Chimney

purging had already been done by air passing through during the 'off' period which had swept out any possible gases and had changed the boiler gas volume over and over again. Even on the plants where purging could be said to be essential, the period was often excessive. Several plants were found to have purging periods of over 30 seconds, and this could be reduced to around 10 seccnds and still give several volume changes. Modern plants have a post-firing purging period which makes the position even more ridiculous if air is permitted to pass unhindered through the plant during the 'off' period.

Early chimneys erected to serve Economic and the earliest packaged boilers were usually of mild steel with no insulation, although the outsides were possibly painted with aluminium heat resisting paint 'to prevent corrosion'. It was soon found that as boiler efficiencies increased these chimneys had extremely short lives, and although aluminium cladding, with an air gap, cured many plants, as efficiencies increased and exit gas temperatures dropped further, troubles recurred and today no firm should instal a new stack with only this degree of insulation. It must be realised that it is not the average gas temperature in the flue or chimney that is an indication whether corrosion or smutting may occur, but the internal wall temperature. Draught stabilisers, widely fitted in earlier installations, admit large quantities of cold air into flues and chimneys and often prevent the internal surfaces of even well-insulated chimneys staying above acid dew point. These stabilisers should be secured in the closed position and retained only as explosion relief doors. Good burners do not need them.

As more research has been carried out, the factors governing chimney design have become more clearly identified, and in turn greater safety margins are being requested. While at first sight the stability of a chimney may not be related to fuel economy, collapse or urgent repair of this item will close down the boiler plant it serves and the loss of profit or production will certainly be felt.

The chimney is affected by two problems, first corrosion both internal and external, and second wind forces which may topple it or set up such stresses that damage by fatigue occurs. The two documents worth studying are British Standard 4076 and Code of Practice 3. Both have been revised frequently, CP3 as recently as 1972. The occurrence of oscillation in steel chimneys has been found to be much more widespread than at first thought.

EXAMPLES OF CHIMNEY TROUBLES

A large brewery had a new steel chimney in service for about five years when it was noticed, actually by a passer-by, that corrosion was causing holes to appear in the metal collar surrounding the top of the chimney. The Brewery were persuaded that the top should be examined in case pieces of the collar might fall, and indeed when steeplejacks examined the top, the metal was found to be badly corroded and the collar would probably have fallen in the next high wind period. However, and only because the ladders had been erected in any case, the opportunity was taken to strip the aluminium cladding from the first joint ring below the top of the chimney to see if any corrosion was evident. In fact there was none, but half the flange bolts were broken in two due to excessive stresses caused by oscillation. The pieces of the bolts were in some cases only retained by the cladding, and it was clear that this fortuitous examination was only just in time to prevent the major disaster of the top 6 to 7 metres of chimney crashing down into the brewery.

The examination continued and the next joint flange was found also to have broken bolts, but only a relative few. The cure was to fit stronger bolts and anti-vibration spirals or strakes to the top external protion of the chimney. The whole point is that when the chimney was designed it conformed to good practice at that time.

Even with apparently well-insulated chimneys trouble can occur. Diagram No. 10 shows a typical construction of twin metal layers with insulation between. Chimneys have usually to be flanged for site erection and in this case the inner metal sheet was brought out as a flange to be sandwiched between the flanges of the outer shell. Although asbestos joints were fitted to try to insulate the inner flanges, trouble occurred and the chimney came dangerously near to collapse. Enough heat was conducted along the inner flanges to allow the metal surface temperature in the 'V' of the inner skin joint to drop below acid dew point, corrosion occurred and after about three years, penetrated completely through to form several small holes. Condensation then penetrated into the insulation, to the outer metal shell and flange rings, and when the chimney was inspected had so weakened it that the upper sections had to be removed for repair. Upon re-erection the flange joints were covered with a layer of insulation protected by a 'bulged' aluminium ring, and it is expected this should prevent any recurrence.

Further examples of fatigue and corrosion are known, and it is recommended that firms and local authorities should have steel chimneys, even those only installed a few years ago, checked against the very latest recommended design standards and the chimneys should also be examined by competent steeplejacks for corrosion and any signs of fatigue cracking, particularly at flanges. Where double-skinned chimneys exist, which of course cannot easily be inspected along internal joints, then small test holes should be drilled to allow the thickness and strength of the internal skin to be checked. Extra insulation, suitably weatherproofed should be added where this seems advisable.

HEAT LOSSES FROM BOILER PLANTS

As has already been illustrated (Diagrams 3 and 4 of Section 1), irrespective of the type of boiler in use, the major loss is the heat carried away in the exist flue gases. The temperature of these gases is primarily dependent on the design of boiler and secondly on fouling of the heat transfer surfaces due to scale on the water side and soot or dust deposits on the gas side. The volume of exit gases at any steady temperature depends on the fuel/air ratio and poor combustuion will lead to increased costs. When oil-fired 'packaged' boilers were first introduced, and existing boilers converted to oil-firing, attempts to operate with low excess air figures (high CO_2 or low O_2

39

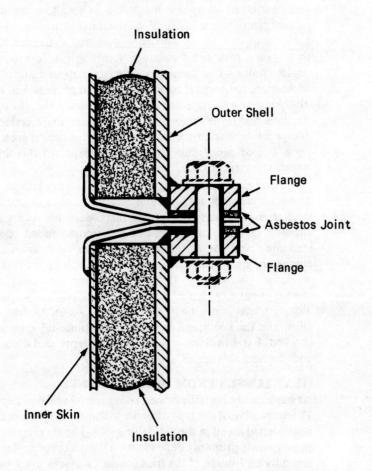

Diagram 10 — Typical Form of Chimney Flanged Joint of Double-Skin Insulated Construction

readings on test instruments) caused carbon deposits and rapid fouling of smoke tubes, so that it became accepted that 10 to 11% CO_2 readings in the combustion products was a good compromise. However design of oil burners has improved, and such readings should no longer be 'accepted'.

Diagram No. 11 shows for the three usual fuels the sensible heat losses at various CO_2 readings, assuming a 245°C (440°F) difference between gas temperature and ambient air temperature i.e. gases at 260°C (500°F) with 15°C (60°F) air temperature. This is drawn on a nett calorific value basis since boilers do not usually condense the water vapour in the gases - indeed if they did, corrosion and acid smuts will be unacceptable on most plants! This diagram is really only absolutely accurate for one ultimate analysis of each fuel, but values have been chosen typical of boiler fuels, and the real value of the diagram is to try to encourage regular checking of combustion conditions so that corrective action can save fuel.

Simple gas temperature indicators are available, and providing elementary precautions are taken to position the sensing portion in the centre of the flue and not too close to any water-cooled surface, the errors are not likely to be too great at the level of exit gas temperatures to be expected from reasonabley efficient boilers.

Flue gas analysers are also widely available, ranging from simple hand-operated chemical absorption types up to very expensive continous recording instruments, which latter can really only be justified on boilers burning considerable quantities of fuel, say over £5000 costs per year.

Boiler operators and/or factory service engineers should be encouraged by management to make periodic checks on their boilers and record the results. Each reading should, if possible, be related to steam output or burning rate and taken during a relatively stable period of output. With oil or gas burners the modulating position of the burner controller is a good guide and with coal the feed rate of the stoker can be used.

Now a few examples (oil is taken as the fuel, but the remarks apply equally to coal and gas).

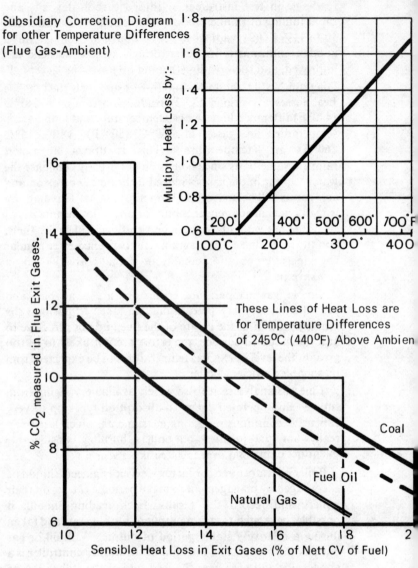

Subsidiary Correction Diagram
for other Temperature Differences
(Flue Gas-Ambient)

Multiply Heat Loss by:-

These Lines of Heat Loss are
for Temperature Differences
of 245°C (440°F) Above Ambient

Coal

Fuel Oil

Natural Gas

% CO_2 measured in Flue Exit Gases.

Sensible Heat Loss in Exit Gases (% of Nett CV of Fuel)

Diagram 11 - Effect of Fuel/Air Ratio on
Heat Loss in Exit Gases from Boiler Plant

1   Taking readings of temperature and CO_2 of the exit gases, it was noted that over a period of 2 months the exit gas temperature at the same load and the same CO_2 reading had risen by 30°C (about 55°F). The reason must be either scale or fouling; thorough cleaning of the smoke tubes reduced the temperature to the original level so the trouble was fouling. This represents a multiplying factor of 1.17%, on the original heat loss of 13.4% a 2.3% drop in efficiency leading to a 3% increase in fuel consumption, value £840 if over a full year There is probably a case for more frequent manual cleaning, or the installation of automatic soot-blowing equipment.

(in these examples a common case is taken of a single boiler operating on a average load of 2 tonnes steam per hour (4400lbs/hr) for 2000hrs per year, using fuel oil at £90/ton so that the annual fuel bill is £28,000).

2   Although exit temperature remained more or less constant, CO_2 figures dropped from 11% to 9%. This increases gas loss from 13.4% to 16.1%, leading to a drop in efficiency of 2.7%, an increased fuel consumption of 3.6% or £1,000 in a full year. As the burner has been demonstrated to be capable of operation at 11% CO_2, no time should be lost in adjusting it, or replacing worn parts etc.

3   The burner on a boiler can only operate at 10% CO_2, no matter what efforts are made by the firm or the burner service engineer. A new burner is on the market with a known performance of 12.5% CO_2 on this type of boiler. The diagram shows that assuming temperatures do not alter appreciably, the heat loss in exit gases would drop from 14.6% to 11.6%; a 3% increase in efficiency or a reduction in fuel consumption of around 4%, worth £1,100 in a full year. If the cost of the burner plus installation is less than say £2,000 this would be a good financial transaction.

With older boilers, such as Lancashire and two-pass Economic types, 'economisers' were commonly fitted to preheat boiler feed water or process water by further cooling of the gases leaving the boiler. Considerable savings could be made as these boilers often had exit gas temperatures exceeding 400°C (730°F) and the gases could be cooled to say 200°C (390°F) without too much corrosion troubles, particularly as most of these plants were coal-fired. There has been a tendency to assume that modern boilers have such low exit gas temperatures that there is no case for such heat recovery devices. Certainly the use of ordinary steel or cast-iron tubes would result in very rapid corrosion with fuels containing appreciable sulphur percentages. However research has produced, for other purposes, many materials much more resistant to corrosion than the traditional materials of the old 'economisers'. One could also consider replaceable units in which the heat exchange surfaces were easily removable and cheap enough to give a satisfactory return on investment if they only lasted four or five years.

Fibreglass flues, coated induced draught fan casings, and chimneys have limitations on maximum temperature, which are sometimes considered perilously near the exit gas temperatures of modern boilers. As a result their use has been rather limited, for fear that scale or fouling in the boiler could cause the exit temperatures of the gases to rise and damage the fibreglass items. However if lower gas exit temperatures could be guaranteed by the use of a heat exchanger, the use of fibreglass or other corrosion free materials could be extended to make the acid dew point corrosion problems become negligible.

For example, if the typical boiler of the diagram, and the three previous examples, had its exit gas temperature reduced from 260°C (i.e. 245° above an ambient of 15°C) to 200°C, the reduction in exit gas loss would be 2.3%, representing a 3% fuel saving. This would save £841 in a full year. Of course such a heat recovery device could preheat air for combustion and give a similar fuel saving if there were no scope for water heating.

WHY BE SATISFIED WITH 80% EFFICIENCY?

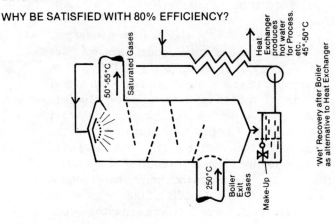

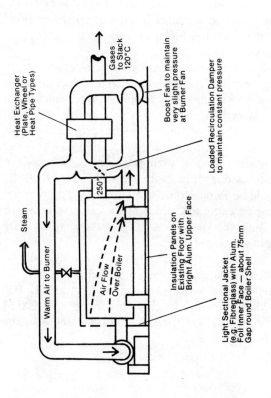

A well-run boiler may have a 2 to 3% radiation loss from its outer surfaces, and a heat loss in exit products of combustion of 41%, i.e. 250°C and 9% CO_2 when burning natural gas. Assuming a further 1 to 2% loss in boiler blowdown this would result in a 'good' efficiency of 81 to 82%. When burning natural gas there are no longer any sulphur acid corrosion problems if the exit gases are cooled even further. Diagram 11A shows two possible methods using equipment already on the market. The first method is to preheat the combustion air for the boiler burner, using first a light jacket around the boiler to direct inlet air over the shell to recover much of the radiation. The air can then be directed through a heat exchanger to the burner. Usually a fan is necessary to overcome the slight resistance of ductwork and exchanger to avoid altering the burner fuel/air ratio. The floor below the boiler can be insulated and an efficient heat exchanger fitted to recover heat from the boiler blowdown to preheat feed water or process water.

If there is a continuous demand for warm process water, or if the boiler is only used for space heating, the Diagram shows an alternative recovery method by a spray cooler, which uses a secondary heat exchanger to give large quantities of warm water, which can be used for boiler feed, process, or space heating via larger size radiators than would be used with water at the more normal 90°C flow temperature.

If the boiler is dual-fired, with oil as stand-by fuel, the systems should be by-passed when oil is in use to prevent too much cooling of exit gases which might cause smutting or corrosion. With the spray system, by-passing really only means stopping the water spray.

By either of these methods the overall efficiency could be raised to around 92%, giving a potential fuel saving of 12%. If a boiler delivered an average 5 tonnes/hr of steam (11 000 lbs/hr) for 2000 hrs/year it would burn around = 85 000 of fuel per year even at 82% efficiency, so a 12% saving would be roughly = 10 200 per year.

3
Space Heating of Factories, Offices & Similar Buildings

INTRODUCTION

Sections one and two have indicated that there are many facets to the choice of fuels, boiler plant and ancillaries and that the acceptance of time honoured practices is not always the guarantee of efficient operation. The heating of existing and future buildings for all functions requires that a fresh look be taken into these time honoured practices that have been perpetrated in an era of plenty at comparative low cost.

At this time there is a general voice of opinion that our insulation standards should be improved. This, however, is only part of the solution to achieving economic running costs for the heating of buildings. Much more must be done to thoroughly master the multitude of sources of heat loss that occur in both old and comparatively new buildings.

Traditionally, the Architect has produced a building design and then the Heating Consultant or Contractor has been called in to design a heating installation that will effect a thermal equilibrium as dictated by the design. The advantages of reversing this process in present day circumstances of high interest rates, the need to reduce running and other overhead costs, plus the promise of ever increasing fuel prices are manifest.

Economic Heating Use of Fuel in Existing Buildings

The prime object of a heating installation is to maintain a steady state temperature level inside relative to the function of the building and that the input of quantity of heat shall be in relation to the fluctuation of the external conditions to maintain the required internal prescribed temperature.

Regrettably, all too often the optimum condition is rarely achieved and in consequence, in many cases, this results in a greater consumption of fuel than is really necessary.

Some of the principal heat losses and causes of unsatisfactory performance that occur to create such phenomenon are listed below:—

1. The heating installation has not been correctly balanced. That is to say a heating installation using low or medium pressure hot water as the heating medium has not had the regulation valves suitably set. Sections of the installation are affected and individual heat emitters therefore do not

49

receive the correct allocation of hot water to enable them to perform their design duty. The resultant effect is that some areas will be overheated and others underheated. Very often, in extreme weather, the popular method to overcome this defect is to boost the temperature in the boiler to satisfy the deficient areas with resultant further increase in waste in those areas already overheated. These observations are made upon the premise that initially the installation was correctly designed. Some of the older installations were regrettably designed on the rule of thumb principle and as such are worthy of further examination in view of present day prices of fuel to see whether additional heating surface should be added in "deficient" areas to give a better balance to the system.

2. Many buildings suffer a very high rate of air change and subsequent heat loss and unnecessary fuel usage due to unsealed holes in the structure. Filling of gaps between roof sheeting at ridge and eaves and attention to badly fitted windows and doors can reduce warm air losses, which should not be underestimated, as they can represent a high proportion of fuel usage until properly sealed.

3. Wall finishes can play a big part in achieving acceptable comfort conditions in offices. Present day methods of calculating room heat losses take into account this aspect. For example, the simple expedient of covering existing walls with a good quality lining paper can improve the internal conditions quite appreciably. Unlined upper sheeted sections of factory walls can be improved by insulation or continuation of the brickwork up to eaves level.

4. Many of the older office buildings have ceilings which are higher than the present acceptable norm. Therefore more air is being maintained at room temperature than necessary and fuel is once again wasted. Many methods can be used to reduce the ceiling height and consequently reduce fuel consumption.

5. Surveys conducted by NIFES in buildings where unsatisfactory conditions have been reported have very often shown that the original design was for a building having a

radically different room layout. Subsequent internal structural alterations have then produced unbalanced conditions. These very often lead to the use of supplementary heating, once again increasing unnecessarily the heating costs.

6. The function of industrial properties very often change due to new occupiers. The original heating installation may be totally unsuited to the new use of the premises. It is desirable that potential new owners or tenants take expert opinion on this point to ensure the suitability of an existing plant for their particular business. Such a precaution ensures that the future occupier is in possession of details of possible future capital expenditure and fuel costs prior to settlement of contract. An example of what can happen quite innocently was the case of a multi-storey factory, which was converted to office premises. The ceilings had been properly lined and heights of rooms reduced, as recommended earlier. Several tenants had signed contracts and one potential tenant expressed interest in the Boiler Plant before signing his agreement. The result was that it had missed the notice of the agents and was found tucked away in accommodation separated from the basement area. It was, in fact, a rather ancient and inefficient manually operated steam boiler plant requiring the employment of two boiler attendants and the use, at that time, of a very expensive form of fuel. Realising their oversight, the owners of the property had plant installed which in terms of saving paid for itself in under two years.

7. Many of the older heating installations were designed on the thermosyphon, i.e. natural circulation principle, necessitating the fixing of pipes at high level near the ceiling or in the apex section of the roof. Circulating pumps were fitted later, but the high level pipes remained. In most cases these pipes now represent an unnecessary and wasteful heat loss and expert advice upon their removal or insulation is advised.

8. If vehicles have to enter the building frequently to load or unload, the large access doors should be "air-locked" if

possible so that one of the two sets of doors can always be kept closed while vehicles pass through. Some buildings have existing double doors, or even have loading bays inside the building so that the door can be shut after a vehicle has entered, but with the increasing size of vehicles the doors can no longer be closed and have to remain open all the time the vehicle is inside. Such loading bays should be rebuilt to accommodate today's larger lorries and vans.

THE HEATING OF FUTURE BUILDINGS

The custom for the Architect to design a building, then enlist the aid of a Consultant or Contractor to handle the requirements for heating is well established. For future buildings, it is recommended that both parties commence co-ordinating their designs at the very beginning. In this way the most economic building design can usually be achieved and full recognition taken of the factors necessary for incorporating in the design to give economy of maintenance and operating costs for the future. The following factors serve to illustrate this point:

Orientation of the Building

On a particularly exposed site it may be advantageous to have one of the shorter sides of a rectangular building facing in the direction of the cooler prevailing wind and also possibly minimise glazing to reduce heat losses, consequently fuel costs. If it is preferred to have appreciable areas of glazing in the facades, then careful orientation can minimise the necessity for comfort cooling in summer for the efficient working of the occupants during periods of high heat gain, due to solar radiation.

The Shape of the Building

It is appreciated that certain manufacturing processes may require a building, which has a wall length to floor area ratio, which is both expensive initially and also by nature of this fact equally expensive to heat. However, careful consideration of building shape can be a means of reducing both initial building costs and eventual running costs. The following example shown in diagram 12 and 13 illustrates this point.

52

Diagram 12 EXAMPLE OF INFLUENCE OF BUILDING SHAPE IN RELATION TO STRUCTURAL AND HEATING COSTS

TYPICAL FLOOR REQUIREMENT FOR 80 PERSONS USING 'A' MULTI-OFFICE AND 'B' OPEN PLAN SYSTEM

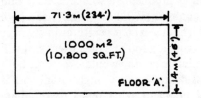
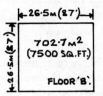

MULTI-OFFICE SYSTEM
PLAN A

9.29m² (100 sq. ft.) REQUIRED/
PERSON ASSUMED 75%
USEABLE

80 PERSONS × 9.29 m²
= 743 m²
= 75% USEABLE

... TOTAL FLOOR AREA
= 743 × 100
       75

= 990 m²

= 71.3 × 14

TOTAL PERIMETER
= 170 m² (560 ft.)

OPEN PLAN OFFICE SYSTEM
PLAN B

7.43 m² (80 sq. ft.) REQUIRED/
PERSON ASSUMED 85%
USEABLE

80 PERSONS × 7.43 m²
= 594 m²
= 85% USEABLE

... TOTAL FLOOR AREA
= 594 × 100
       85

= 700 m²

= 26.5 × 26.5

TOTAL PERIMETER
= 106 m (384 ft.)

Diagram 13 COMPARISON OF HEAT LOSSES OF DIFFERING SHAPED BUILDINGS

BUILDING HAVING 3 No. TYPE 'A' SIZE FLOOR HEATLOSS WATTS/°C (BTU°F) TEMP. DIFF.			BUILDING HAVING 3 No. TYPE 'B' SIZE FLOOR HEATLOSS WATTS/°C (BTU/°F) TEMP. DIFF.		
	Watts/°C	(BTU/°F)		Watts/°C	(BTU/°F)
ROOF	791	(2,700)	ROOF	553	(1,888)
FLOOR	190	(648)	FLOOR	132	(453)
WALLS	399	(1,362)	WALLS	426	(1,455)
GLASS	2,435	(8,310)	GLASS	532	(1,818)
TOTAL	3,815	(13,020)	TOTAL	1,643	(5,614)

ASSUMING 15% GLASS OF TOTAL WALL AREA

NOTE: BUILDING 'A' HAS 44% MORE FLOOR AREA THAN 'B' BUT HEAT LOSS IS 130% GREATER!

The Requirement of Windows

In both commercial and industrial buildings the examination of the necessity for glazing should be carefully considered. Large areas of single glazing constitute a very high heat loss factor in windows as well as producing a source of heat gain from solar radiation in summer. High heat losses mean high fuel costs in winter and high internal temperatures in summer mean discomfort to the occupants and loss of efficiency. It will probably be argued that double glazing would overcome both problems. However, consideration of the high initial cost is essential and in general double glazing tends to reradiate internal heat gains in summer, so increasing rather than diminishing the problem.

Glazing is desirable in certain buildings for psychological reasons and the percentage, positioning and shape of glazing of the wall area requires careful consideration.

Materials used in Building Construction

The process of heating a building up from cold, the time taken to reach the desired internal temperature level and the amount of fuel necessary are related to the materials used in its construction. Every building has a thermal capacity, that is its ability to absorb heat put into it. A building constructed of materials having a low thermal capacity will, therefore, take less time and fuel to heat up than one having materials having a greater geat absorbing capacity. Careful choice of building materials should, therefore be taken, at the same time recognising the structural strength of the materials in relation to the particular building. It is possible that the initial cost of the low thermal capacity building may be higher than that of one having a higher rating. However, the difference in cost should be set against the resultant saving in fuel costs. Such fuel costs should include an escalation factor. Expertise in the use of low thermal capacity materials in conjunction with those having a high rating can be exercised, so that a compromise can be achieved to produce a building of reasonable capital cost, but having a minimum thermal capacity.

SPACE HEATING AND OCCUPANTS COMFORT

Considerations of Occupancy and Manufacturing Patterns

The choice of the type of space heating should be in

relation to the type of activity carried on in each particular area of the building envelope. In the interest of fuel economy and running costs it is, therefore, desirable to examine at design stage the pattern of occupancy in relation to the function of the building and also the pattern of the manufacturing or construction processes carried on within the building. The following are two simple examples, however, it is emphasized that each individual case should be assessed carefully.

A factory may be engaged upon the manufacture or servicing of equipment having a high metal content. Such equipment may have a high thermal capacity and if the throughput of such equipment is fairly rapid, it can be appreciated that this equipment will tend to depress the overall temperature level of the factory and also increase the rate of heat input necessary to maintain satisfactory conditions. In such circumstances structural design of the building could be possibly so arranged as to minimise or totally eliminate this absorbance of heat by the equipment, the economics of such a proposal being assessed at the design stage of the building.

A part of a building may have a number of rooms, which are regularly only used and require heat for a small percentage of the day. It may, therefore, be worthwhile initially increasing the degree of insulation to that particular section to achieve a limited heat up period prior to occupancy, the increased cost of the insulation being offset by the reduced period that heat will be supplied. Careful initial study and a full understanding of the patterns of use of a building can produce a policy of minimum fuel consumption. In certain circumstances such a study can indicate that use of two or more heating media may be economically the more acceptable.

Comfort of Occupants

The heat which is required to provide acceptable comfort conditions to any area of a building is to some degree related to the degree and type of activity of the occupants. Reference has been made earlier to the papering of existing plaster or paint finished walls to reduce the sensation of discomfort to a sedentary worker seated in close proximity to such a surface.

55

The equivalent temperature demanded by the human body depends upon the air temperature around it, radiation received by it, radiation from it to cold walls and windows. These factors influence the heat loss from skin and clothing. Appreciable air movement occasioned by ill-fitting windows and doors also increases the sense of discomfort.

Generally an equivalent temperature level between 18°C (65°F) and 21°C (70°F) will be acceptable for the less strenuous activity situations where work is of a sedentary nature. If room air is below the equivalent temperature level for body comfort, it is necessary to compensate for this difference by radiation. Therefore, walls, floors and ceilings and inter-mediate surrounding surfaces should be slightly above the air temperature. Cold surfaces produce discomfort and a demand for higher air temperatures. Careful consideration of wall, floor, ceiling and surrounding finishes is desirable. To reduce the incidence of hot heads and cold feet, the reverse, or at least a reasonably small temperature gradient should exist, since, the average person prefers to have the feet warmer than the head. However, as this situation is normally only applic-able where the floor is heated, careful design and siting of heat emitters is the best practical solution. It is essential to ensure that the temperature at floor level is not less than 1.8°C (3°F) below that at head level. Discomfort is usually felt with floor temperatures in excess of 24°C (75°F).

For occupants having a greater degree of activity, a lower air temperature may be desirable, particularly in instances where it is necessary to wear protective clothing, which acts as an insulation against those body losses due to air movement and cold surrounding surfaces.

These situations must be assessed individually with the possibility of 'spot' heating of areas to reduce heating costs. It is important to remember that even slight overheating is very wasteful. In Britain, over the whole heating season, the outdoor temperature averages around 7.2°C (45°F). Naturally this varies slightly from area to area, and in other countries may be considerably different and if desired actual values can be found in reference books, e.g. the I.B.S. "Guide to Current Practice". If a building is maintained at 19.4°C (67°F) instead of 18.3°C (65°F) when this latter temperature is known to be satisfactory, then the temperature differential is 10% greater (i.e. 22°F instead of 20°F). The fuel consumption will then be 10% greater also.

METHODS OF SPACE HEATING

The wide choice of equipment available for space heat today necessitates that careful selection be made in order to obtain the best system at an economic cost and reasonable operational cost.

CONVECTION HEATING

Radiator Heating (Heating Medium – Low Pressure Hot Water)

Steel radiators have largely replaced the cast iron type, and being of more attractive appearance are more acceptable, particularly in the domestic market. The name radiator is a misnomer, as over 75% of the heat output is by convection (the movement of air occasioned by heating, over the hot surface). Radiators are most economically employed in smaller rooms giving the advantage of individual room control, either by hand or thermostatic valve. Radiators used in larger areas, although popular because of low initial cost, give rise to high temperature gradients in higher than average rooms. This is wasteful fuel consumption, even if there is a floor above to absorb some of the heat.

Natural Convector Heaters

Heating Media – Low pressure hot water; medium pressure hot water; steam. Perform a similar function to radiators, can be individually controlled, and are best used in smaller areas of a building. The heating element is normally a tubular header with fins attached. In order to maintain output, the elements should be kept clean using a vacuum cleaner. Heat transfer is almost wholly by convection with small radiation component.

Fan Convectors

Heating Media – Low pressure hot water; medium pressure hot water; steam. They have a high output, a wide range of application in the larger areas in schools, hospitals and commercial buildings, have the advantage of being able

57

to operate automatically, give elevated output of 'boost' for quick heating and some lower settings under thermostatic control.

Units usually have air filters incorporated. To maintain performance they should be cleaned at least three times per heating season. The cost of this operation will probably now be £5.00/unit/year/. Economic in running cost if central override controls are provided to eliminate waste with external temperatures at 15°C (60°F).

Unit Heaters

Heating Media – Low pressure hot water; medium pressure hot water; steam. Basically a propeller fan and finned heater battery in a metal casing. They have a high output and can be used to advantage to give clear floor areas and in large factory areas, normally fitted at high level with horizontal or vertical air discharge. Careful choice has to be made in selection to give necessary 'throw' to cover floor area in relation to discharge temperature of air and mounting height. For this reason use on low pressure hot water is limited. On/Off operation of fans through thermostats. The system can be wasteful in consumption of fuel if high level supply mains are extensive and left uninsulated.

RADIANT HEATING

Commercial Heating

Heating Media – Low pressure hot water; medium pressure hot water. Proprietary systems of room ceiling mounted radiant heating using a small bore pipe mounted above a metal tray system have wide acceptance for clean appearance, close temperature control and the advantage of clear floor and wall spaces. Control of unnecessary thermal gain to structural ceilings above the coils is desirable to restrict operation costs.

Industrial Heating Panels

Heating Media – Medium pressure hot water; high pressure hot water; steam. Radiant panels used in industrial buildings to heat large factory areas. Panels usually have single sided radiant face mounted horizontally or inclined. The radiant 'cone' of each is designed to heat the floor and parts of the walls and objects being handled without effecting a super elevation of the air as the intermediate medium. Such systems are advantageous in heavy engineering workshops where high bay factories are necessary. System reduces temperature gradient normally associated with high roof heights and also reduces fuel consumption. Time lag is minimal due to immediate radiation effect which makes them attractive for intermittent type heating. Radiant panels are acceptable for 'spot' heating of small areas leaving the remainder of the building unheated.

Floor Heating Panels

Floor heating panels using small bore hot water pipe coils have had wide use in the past. Difficulties of thermostatic control during rapid weather changes because of high heat storage capacity of the floor make this system somewhat less attractive. The necessity to limit floor surface temperatures to less than 80°F normally necessitates all floor area to be covered by coils. This type of system is generally less attractive financially compared to other alternatives available.

Fan Assisted Warm Air Heating

Several alternative systems are available using free standing heaters with or without distribution ductwork. Heat exchangers use either gas or light oil burners as a heat source. Thermostat control is effected by a sensing device fitted in an averaging position. The ability of these units to cover large floor areas makes them economic in initial cost. Fuel consumption can be higher than other systems unless there is an elaborate system of air distribution. Even temperature control can be obtained by the use of supply and extract ductwork, but this increases initial cost to a less attractive level. Convection heating requires a greater heat input than with radiant heating. Tests have shown that 'radiant' heating may use 15% less fuel than a 'convection' system to give the same degree of comfort.

Combined Heating and Ventilation

Architectural design over the past 25 years has tended to incorporate large areas of glazing giving rise to high summer solar heat gains and also high heat losses in winter. The absence of air conditioning in these buildings caused a measure of discomfort to occupants due to high internal temperatures in summer. In winter due to the large areas of glazing being at almost external temperature, the chilling effect on air due to the current of cold air passing down over the face of the glazing caused cold feet and general body radiation loss due to the cold glazing. To improve the summer conditions local extract fans were provided in windows or a fan assisted ducted extract system with extract grilles in the ceiling or inner walls was provided. Extracted air being counter balanced either by opening windows or the provision of a fan assisted ducted air input system. The need to minimise noise from external sources made the opening of windows undesirable and where double glazing was provided problems were experienced in installing fans in windows.

Buildings having a depth of 20′ or more from the external wall demanded that artificial lighting be used for quite a large part of the working day. The trend to increase the artificial lighting intensities in recent years further added to the overheating problem.

The resultant concentration of both lighting fittings together with supply and extract ventilation grilles in the ceiling created problems of integration and maintenance. Thought has therefore been given to tidying up the ceiling and also the necessity of utilising the heat produced by the lighting fittings. The Integrated Environmental Design Concept was pioneered by the Electricity Authority and a typical layout indicated on Dia. 14. There are numerous alternatives to this basic scheme which utilises the heat extracted from combined lighting and air extract fittings and redistributes it through the air handling plant to thermostatically controlled zones on each floor level. It can be seen that the perimeter curtain wall effect over the external walls is provided to overcome the problem of body radiation loss.

This system has very wide application in multi-storey buildings due to its economic use of fuel by the recycling of heat from lighting fittings. The resultant temperature control is that internal conditions can be maintained at a constant level in summer and winter. In extreme conditions heat will be required for topping up the heat requirement but this is

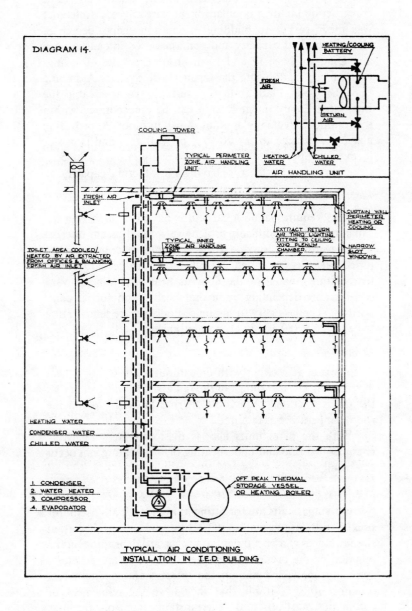

DIAGRAM 14.

HEATING/COOLING BATTERY

FRESH AIR

RETURN AIR

HEATING WATER

CHILLED WATER

AIR HANDLING UNIT

COOLING TOWER

TYPICAL PERIMETER ZONE AIR HANDLING UNIT

FRESH AIR INLET

CURTAIN WALL PERIMETER HEATING OR COOLING

TOILET AREA COOLED/HEATED BY AIR EXTRACTED FROM OFFICES & BALANCING FRESH AIR INLET

TYPICAL INNER ZONE AIR HANDLING UNIT

EXTRACT RETURN AIR THRO LIGHTING FITTING TO CEILING VOID PLENUM CHAMBER.

NARROW SLOT WINDOWS

HEATING WATER
CONDENSER WATER
CHILLED WATER

1. CONDENSER
2. WATER HEATER
3. COMPRESSOR
4. EVAPORATOR

OFF PEAK THERMAL STORAGE VESSEL OR HEATING BOILER

TYPICAL AIR CONDITIONING INSTALLATION IN I.E.D. BUILDING

61

considerably less than that used in the traditional frame construction building popular for the past 25 years. Reference was made earlier to building orientation shape, degree of glazing, thermal capacity and materials used in construction and Diagram Nos. 12 and 13 illustrate examples of savings that can be made using the Open Office System. Assuming that the Open Plan System is not always acceptable the Integrated Environment Design can be incorporated within the traditional rectangular plant shown in Floor 'A' providing glazing is reduced to say 20-15% of wall area and is in slot form with appropriate attention to materials used in construction to form a structure having a low thermal capacity.

Electric Underfloor Heating

This requires a floor structure having a high thermal capacity and it is difficult to achieve a steady state internal temperature condition due to fluctuation of external temperatures. Heat is provided at off peak tariff charges and temperature 'swing' of the floor as heat reservoir can vary considerably depending upon the specification for its composition. The necessity to minimise floor surface temperature to below 26°C (80°F) has been stated earlier.

Electric Block Storage Heaters

These can be either with or without a fan to boost heat dissipation. The advantage of the fan assisted units being that the fan is operated by a room thermostat and tops up the room temperature as required.

With the plain units heat is dissipated at a relatively uncontrolled rate and underheating or overheating can occur as a result.

Gas Fired Radiant Panel Heating

Individual units are for fitting at high level, are useful for smaller installations and 'spot' heating applications; in certain circumstances can be a potential fire hazard depending on the function of the premises so heated. The units are generally manually controlled and in the interest of maximum fuel economy it is desirable that they have the refinement of automatic start-up and thermostatic control facilities incorporated.

AUTOMATIC CONTROL SYSTEMS

Many existing heating installations have automatic start and stop controls for the daily cycle of operation and also a quite sophisticated system of thermostatic control.

Regrettably, due to the degree of sophistication, many of these controls do not perform their designated duties correctly.

This is due mainly to a number of reasons:—

(a) The apparatus has never been set-up properly.

(b) Subsequent visitors to the premises have 'fiddled' with knobs, without having the requisite knowledge to 'play the correct tune'.

(c) The staff in charge of the plant, very often the caretaker, has never been instructed or cannot really understand the technicalities.

(d) Staff changes occur and new staff are not correctly trained.

It is imperative that all existing controls are made to operate correctly before pronouncing that an installation is not operationally efficient. The simplest form of thermostatic control is possible the room thermostat that switches the heating pump on and off to control the temperature of the small domestic heating installation. These thermostats have to be sited in a position, very often in the entrance hall, to control the temperature to give the designed optimum in the rooms of the dwelling, a very elementary form of zone control. The degree of overshoot or otherwise is not greatly significant in this case and can easily be detected and quickly adjusted by the owner of the dwelling. In industrial and commercial buildings the facility for noting overshoot is possibly non-existent and the consequential heat wastage is many times that of the example above.

Like other heating equipment, there is a wide choice of thermostatic controls available to select from. It is, therefore, essential to accept two simple rules when choosing equipment:—

1. The system shall be as simple as possible to perform the desired function.

2. The proposed expenditure shall be in relation to all the factors appertaining to the mode of operation, running, maintenance and attendance costs.

Simplicity on smaller installations is desirable, so that less experienced engineers can check and rectify settings on thermostats and other controls if necessary, time and money being saved by not having to await the visit of a specialist control engineer from the manufacturers.

The expenditure on the larger installations has to be justified so that to achieve the desired results you purchase the optimum and not the ultimate, the latter may contain some facilities which really are not worth the expenditure in relation to their ability to save on heating costs.

Generally for space heating thermostat controls can be placed into two categories:—

(a) Master Central controls.

(b) Intermediate or zone controls.

Master controls generally consist of apparatus located in the boilerhouse or other heat source. This control usually involves a mixing valve (to blend water from the boiler and that returned from the heating system) which operates under the control of an external detector sensing the external temperature conditions and sometimes taking into account the influence of wind, rain and possibly solar radiation to allow the correct temperature of hot water to flow to the heat emitters to maintain a thermal balance internally. This type of control can work very satisfactorily when there is a common air temperature around the building. Conditions cannot always be satisfactory when one or more elevations of the building are subject to differing external temperatures, due to the incidence of solar heat gain. The external detector is normally placed upon a Northerly facing wall so that its sight of the sun is minimal. The situation can arise where the detector is calling for heat for all areas whereas there are areas in the line of the sun's rays which are now satisfied due to this temporary additional heat source. In order to effect a saving upon fuel usage, employment of an intermediate or zone control to restrict the supply of heat to those areas receiving this temporary heat gain is desirable.

Zone controls to complement the master controls can be the means of both saving on fuel costs and assisting in staff recruitment and retaining staff. Staff are much more discerning today on the subject of working conditions in buildings and uncomfortably high temperature zones are to be avoided wherever possible.

Zone controls in the form of a temperature sensing detector operating a diverting valve will simply reduce the heat input into that section of a building receiving in addition other uncontrolled heat input from either solar radiation, electric lighting or electrical calculating machines.

When heat producing equipment is added to either a factory or office area, recognition of its potential to alter the environment should be made so that appropriate action is taken beforehand rather than later. Many heating installations have circulation losses that could be saved if appropriate action were taken. For example, schools are now extensively heated by fan convectors. In the between season periods, March to April and September to October, temperature at 10 a.m. is quite low, however possibly 2 hours later the sun may have increased that level to 60°F. At this external condition it is recognised that no heating is normally required to a building. It has been found that there is very often a blind spot in control systems at this period with windows being opened and consequently fan convector heaters have been influenced by intake of cooler air and continue to emit heat unnecessarily. In such conditions a master override facility at the boiler, controlled by an external sensing element will not only eliminate this but also save the distribution pipework losses by isolating the flow of hot water from the boiler. Output from uninsulated pipes and from the heaters even with fans stopped can often be over 10% of design output so causing overheating with mild external temperatures. The recent introduction of a new form of control complementary to the Master system is the optimum start time control of heating installations. It has produced claims of major savings in fuel costs and, as an interim effort to limit waste of fuel, is to be commended. Basically the heating system is switched on at a time each day which is variable in relation to the external temperature condition. Earlier systems switched the apparatus on at a fixed predetermined time each day regardless of external temperature, which had the possible disadvantage of either underheating or overheating, dependent upon external conditions. Even with individual room thermostatic control, the building was often held at design temperature for more hours per year than justified by the hours of occupancy. Many of the existing controls used on heating installations can be augmented to give additional fuel savings and expert advice should be sought in this direction.

The observations made relative to thermostatic controls may appear to be over simplified. Systems must be broken down into their elements so that their purpose can be understood by the non-technical mind and enable detection of a fault in their operation, to allow calling-in of someone capable of correcting the fault.

CONTINUOUS OR INTERMITTENT HEAT AND BOILER MARGIN

Calculations of heat losses for buildings are based upon providing the heat necessary to match the steady state loss condition and the design base of say $-2°C$ ($30°F$) external temperature. This does not cater therefore for the heat to overcome initially the thermal capacity of the building referred to earlier. Use of intermittent heating, i.e. for the 5 day working week, requires this to be overcome. The system should, therefore, have a reserve, this reserve is in the form of a margin over nett heat requirement of the building and is in the form of additional boiler potential heating capacity.

It can be argued that with continuous heating a boiler margin should not be necessary, however, such a reserve is for quick heating from cold after a breakdown and further protection on the few days each year of external conditions below –2°C (30°F). Continuous heating at a reduced temperature, say 10°C (50°F), when buildings are unoccupied or out of working hours produces only marginal savings and intermittent heating is the best overall seasonal policy. Boiler margins should be correctly calculated as variations can differ quite widely dependent upon building construction, figures of 40%-60% are quite common. The ideal factory for single-shift heating would be a lightly framed one, sheet covered and internally insulated to give the shortest possible preheating period. Radiant type heaters may give a quicker feeling of comfort than convector units.

THERMAL INSULATION

Economics

The importance of thermal insulation in the structure of a building has been emphasised earlier in this section, insulation being required for preservation of the temperature in the controlled zone and to prevent wastage of fuel by the unnecessary dissipation of heat. Of course although this Section is primarily concerned with space heating and buildings, better insulation is equally desirable for pipework and vessels.

The degree to which thermal insulation should be extended is normally dictated by physical requirements, but largely by the balance between the cost of applying and maintaining the insulation and the value by way of the fuel saving involved. With forms of thermal insulation, which are not subject to variations of thickness, e.g. cavity wall insulation and double glazing, economic considerations require the equation of the capital cost and the period of time necessary to recover this cost by savings in fuel cost.

Thermal insulation of vessels and pipework, particularly, require determination of the economic thickness of insulation employing the 'Minimum Cost Method' laid down in BSS.1588: 1963. The determination of two main factors are necessary for the exercise.

(a) The cost of the insulating material per unit length or area.

(b) The value of the heat lost through the insulating material during an assumed period over which it is desired to pay for the insulation out of savings.

Certain components of these factors require more detailed consideration.

1. *Capital Cost*

In the case of a new building or plant, it may well be that the reduction in energy requirements (which may be heating or cooling) arising from the application of insulation will make possible a reduction in the size of the boiler plant, chilling plant, or heating/cooling installation. The cost savings arising should, of course, be deducted from the cost of the insulation.

2. *Write-off Period*

The length of time over which the capital sum should be written-off is arbitrary and depends not only upon the likely life of the building or installation, upon Company Policy, allowance for interest on capital, of high interest rates. Having determined the length of time and the interest rate, the annual sum required to liquidate capital and interest can be calculated from the formula.

$$P = 1 / \left(1 - \dfrac{R}{\dfrac{1}{(1+R/100)^n}}\right)$$

Where P = The percentage proportion of the Capital Sum which has to be paid annually to liquidate the capital and interest.

R = The rate of interest.

n = The term of the fund in years.

THE VALUE OF HEAT LOST

The cost of energy saved by thermal insulation requires that the cost of the fuel itself should be taken into account, adjusted for the thermal efficiency of usage and for special charges related to a particular fuel, such as the cost of pre-heating heavy fuel oil.

Average costs adjusted to late 1979 are as follows:—

Fuel	Typical Price	Possible Usage Efficiency	Cost per useful	
			Therm p	MJ (1) p
Coal	= 36 per ton	74% (3)	18.11	0.172
Heavy Fuel Oil	43.0p/gallon .5p heating (2)			
	43.5p/gallon	77% (3)	31.92	0.303
Light Fuel Oil	55.0p/gallon	77%	43.55	0.413
Natural Gas	21.0p/therm	77%	27.27	0.258
Electricity—on peak	2.5p/kWh	99% (4)	73.97	0.701
Electricity—off peak	1.4o/kWh	99%	41.42	0.393

NOTES:
1. 1 Therm $ 105.5 MJ.
2. This is cost of heating storage and preheating up to nozzle of burner.
3. Assumes good boilerhouse practice, no periods of very low load and allows for idling, weekend losses, etc.
4. Some cable, etc. losses allowed. Remember 2/3 primary fuel loss at power station.

From experience over the last 20 years, inflation should be included in fuel prices. What .annual rate is open to argument, but it is quite clear that although the capital cost is fixed as soon as the work is carried out, the annual fuel saving should *NOT* be taken as fixed, but should be increased each year by some percentage.

THERMAL INSULATION OF BUILDING STRUCTURES

Minimum statutory requirements for the thermal insulation of Industrial Buildings were prescribed in the Thermal Insulation (Industrial Buildings) Act 1957 and relates to buildings which are factories within the terms of the definition contained in Section 151 of the Factories Act 1937. The Act covers only the roofs of buildings and in brief requires that the 'U' value (heat loss) of the roof shall not exceed 1.7 W/m²°C (0.3 Btu/Ft²h °F) at typical heating conditions of 21°C (70°F) inside temperature. Recent large increases in fuel prices necessitate that such standards be appreciably higher than the Act requires. It is highly desirable to ensure that all the elements of a building to be heated are more than adequately insulated, as a hedge against further increases in the cost of fuel.

Traditional methods of insulation include rigid boards of compressed wood or other vegetable fibre, paster and glass fibre. Asbestos can be obtained in board form or applied by spraying. Roofs and floors can be insulated by means of mattresses formed of glass fibres. To these must now be added plastic insulation material in the form of foamed or expanded polymers. These may be built into or injected into wall cavities, ducts and chases as well as providing a surface treatment.

Since the insulating efficiency of the material is largely a function of the air imprisoned within the cellular structure, the thermal conductivity (k) is dependent upon density and varies considerably. A representative value for a foamed plastic derived from Urea-formaldehyde is 0.03 W/m°C. With such a material a wall constructed with a $4\frac{1}{2}''$ brick outer skin, $2''$ cavity, $4\frac{1}{2}''$ brick inner skin, faced with $\frac{3}{4}''$ plaster, uninsulated, has a 'U' value of 1.70 W/m² °C (0.30 Btu/f² h °F) but if the cavity be filled, the 'U' value reduces to 0.45 W/m² °C (0.0792 Btu/f² h °F). The present day price of providing this insulation would be 180 p/m² (150p/yard²) approximately. Assuming such a wall to be heated for 14 hours each day over a period of 39 weeks per year, with an internal temperature of 18.3°C (65°F) the annual heat saving through 1 m² of insulated wall surface would be approximately 191 MJ (1.816 therms). This assumes a load factor based on published degree day data of 0.5 i.e. an average external temperature during the heating season of 7.22°C (45°F). If the heating system be gas fired, the annual saving would be 49.5p or, with light oil, 79.0p. The cost of insulation would thus be saved in 2½ to 3½ years, neglecting interest charges etc.

There is now on the market a wide range of cellular plastics, which are used for insulating material in the form of blocks, board sheets and in liquid form for in situ foaming. The polymers mainly used for the purpose include polystyrene, polyvinyl chloride, urea-formaldehyde, polyurethane, polythylene (polythene) and eborite. Care should be taken to evaluate fire and/or fume risk from certain of these.

Whenever the relative humidity within a heated room becomes high, due to the evaporation of moisture from the person or to processes carried on within the room, there is the possibility of condensation taking place on cold surfaces such as external windows or behind permeable surface coverings, of which insulating materials of most kinds can be included. In addition interstitial condensation can take place within the permeable lining itself.

The possibility of condensation is calculable and if it is likely to occur, should be guarded against, preferably by means of adequate ventilation (sometimes more adequate ventilation of the heated space will do the trick) or by using a moisture barrier. A barrier on the warm face is best, provided that it is complete and does not allow water vapour to creep in round the edges. In such an event the barrier would be useless, since it would imprison moisture within the structure. On some air conditioning applications the vapour pressure gradient may change significantly between day and night and a vapour barrier on each side of the structure would be necessary, together with ventilation of the intermediate structure. It must be admitted that ventilation of cavities leads to increased heat losses. Expertise should be sought to avoid structural damage due to water accumulation.

Another important aspect of building construction is glazing. The use of double as opposed to single glazing is determined not solely by the economic viability, but by other factors, such as reduced condensation and sound attenuation. Reduced condensation is not, however, an unqualified virtue, since the use of a window as a condenser might serve to reduce the more undesirable characteristics of interstitial condensation referred to above. The economics of double glazing should be considered and using the conditions in the example above, we might examine 1 m^2 of glazing. The 'U' value of single glazing may be taken as 5.6 $W/m^2\ °C$ and for double glazing with 12 mm air space 3.3. The annual saving would thus be 351 MJ, and the value with gas firing 91p or with oil firing 145p. Since the cost (unfixed) of clear double glazing units of 6 mm float glass is currently £25/m^2, and the cost of 6 mm single float glass £9 m^2, the difference of 16 would only be recovered in 11 to 17 yrs. Hardly a viable situation, unless one takes a long term view. The addition of interest to capital depreciation would completely eliminate the saving involved. In a new building of course if double glazing allows the capital cost of the heating/air conditioning system to be reduced the case may be better.

TYPICAL EXAMPLE

To illustrate the points made in this Section about heating standards, insulation and air change, consider a small factory building for a typical light engineering activity. Assume this is 30.5 m (100ft) long, 15.2m (50ft) wide, of corrugated asbestos sheet construction with 14% of glass area in the roof and walls, a concrete floor and a cubic capacity of 2830m^3 (100,000ft^3). As originally constructed the normal air changes averaged four per hour, and it was considered that 21°C (70°F) should be maintained inside the building to suit the mainly female labour employed.

In the part of Britain where this factory was situated the average outdoor winter temperature was 6°C (43°F). The building was heated by steam from an existing boiler plant and horizontal discharge unit heaters were to be used. The steam was metered and the **1974** charge was £2.50 per tonne (112p per 1000 lb.) – see Diagram 5 of Section 1. The effective heating hours were 2000 per year.

The following developments took place, or were considered, with all costs and savings quoted at **1974** levels for consistency.

1. *Original building* (constructed before the 1957 Thermal Insulation Act).

   The temperature immediately below the roof was found to be 26.7°C (80°F) due to a fairly high temperature gradient. The unit heaters were mounted about 4 m above the floor and their air input was at this level. The average heat requirement was 192 kW (655,000 Btu/hr), using 350 kg/hr (773 lb/hr) of steam costing £1750 per year.

2. *Modified building* (still uninsulated).

   The unit heaters were modified, vertical inlet trunkings being fitted so that air was pulled down from immediately below the roof ridge. Steam supply and condensate return pipes were insulated to prevent heat rising from them by convection, which helped to cause the original high temperature gradient. All air leakage at sheet joints, particularly at the ridge and eaves, was prevented by using plastic sealer pieces and self-closing rubber swing doors were fitted round the external walls. The temperature immediately below the roof dropped to 23.5°C (74°F) and air changes dropped to two per hour. The average heat requirement dropped to 149 kW (508,000 Btu/hr) using 270 kg/hr (595 lb/hr) of steam costing £1350 per year. Thus a 23% saving was made, value £400 per year, for an expenditure of around £900.

Roof insulated

The roof sheeting was then insulated to the standards required by the 1957 Act using 0.5 inch insulating board with a small air gap between these boards and the asbestos sheeting. The average heat requirement dropped to 98 kW (335,000 Btu/hr) using 177 kg/hr (392 lb/hr) of steam costing £885 per year. This was a further saving of £465 per year for an expenditure of around £2200.

Higher standard of insulation

A layer of glass or mineral wool, or cellular plastic block, could have been inserted between the insulation boards and the asbestos sheeting. This would have halved the loss through the roof. The walls of the building could also have been insulated to this same high standard. This would have reduced the average heat loss to 60 kW (206,000 Btu/hr) and the steam usage to 109 kg/hr (241 lb/hr), costing £545 per year. This further saving of £340 per year could have been justified if the work had been carried out while the roof was being insulated since the major cost of insulation is usually labour, access scaffolding and board retaining clips or channels. The extra work would have cost around £900 if carried out simultaneously with Stage 3, but about £2000 if scaffolding had to be re-erected and all roof sheeting removed while extra insulation inserted.

Double glazing

Consideration was given to double glazing. However this would have only saved a further 5.3kW (18,000 Btu/hr) and the saving of around £50 per year could not possibly be justified, as some 130 m² of glazing was involved which

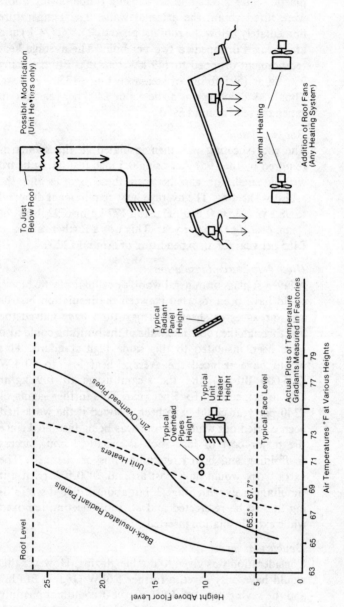

Temperature gradients, with temperatures below the roof much higher than needed at working level, increase heat loss by conduction through roofing and glazing and by air leakage through joints in the roof. Radiant panels usually give the lowest gradient; overhead pipes are the worst. With unit heaters consider inlet trunks to pull air down from under the roof. Alternatively a separate fan system could be installed to force hot air downwards.

74

would have cost around £2000 if double glazing panels had been supplied and fixed in place of the old glass.

Indeed if all stages up to No. 4 had been carried out, the major remaining heat loss from the building would have been the two air changes per hour which would have represented 54% of the total heat losses. These two changes represent the fresh air requirements for over 200 workers where no impurities are generated by the process. In such an industrial building it would be difficult to employ more than one person per 10 m² of floor so that only 50 could comfortably be accommodated. However, considerable expense would be incurred to reduce the air changes in such a building to less than one per hour, involving complete sealing of sheeting joints, air-locking of entrances and possibly a full mechanical ventilation system and such costs would not be economic in view of the savings already made. Such good insulation, of stage 4 standard, is also helpful in summer by delaying the building temperature rise due to solar radiation, keeping the building at a more even temperature over the whole 24 hours of a sunny day and reducing the peak temperatures.

THERMAL INSULATION OF PIPES AND VESSELS

A decision must, in the first place, be made as to whether a particular pipe should be "lagged" or not. Sometimes a pipe serving process plant is left bare on the grounds that it also serves as a heating pipe to the space through which it passes. This is wrong, as it means that the pipe is wasting heat during the summertime. Process steam distribution systems should always be kept separate from the space heating pipework. Insulation of the pipes in the space heating system should depend upon the design of the system itself, and the method of control. For a system of unit heaters controlled by thermostats stopping and starting the fans, pipes should be insulated. Circulation losses are, therefore, eliminated as far as possible. These pipes are usually at high level and do not give much useful heating. Instead they increase the temperature at roof level and increase structural heat losses. Having decided that a pipe shall be insulated, it becomes necessary to determine the extent to which the material shall be applied. The economic thickness is that thickness at which the sum of the insulation cost and the cost of the heat loss over a given period of depreciation is a minimum.

The following example will indicate the operation of the first method, known as the 'Minimum Cost Method'.

Insulation in the form of glass fibre rigid sections with aluminium foil finish is to be applied to a pipe 150 mm (6″) internal diameter carrying steam at 8.274 bar (120 psig) and at saturation temperature 177°C (350°F). The steam is supplied at a cost of 10.4p/therm (0.099p/MJ) based on fuel cost, plus fuel heating only, from an oil fired boiler plant. It is required to determine the economic thickness of insulation related to an operating period of 40,000 hours (say 5 years continuous 7 day operation with annual shutdowns). The following information is obtainable from manufacturer's data, used in conjunction with an estimate of installation costs and calculated energy cost.

INSULATION THICKNESS mm	25	32	38	44	50	63	75	90	100
Cold Face Temp. °C.	45	36	33	31	30	28	26	25	24
Heat Loss Whr/m.	142	120	104	92	84	71	63	56	50
Cost of Heat Loss over 40000 hr £	20.20	17.04	14.80	13.12	11.91	10.11	8.87	7.93	7.10
Capital cost of Insulation £	2.68	3.05	3.44	3.86	4.31	5.90	6.83	7.85	9.00
Total Cost £	22.88	20.09	18.24	16.98	16.22	16.01	15.70	15.78	16.10

It will be seen that the lowest combined total of heat loss cost and capital cost is achieved by the use of insulation 75 mm (3″) thick. The figures given above are shown graphically in Diagram 15. However, if allowance is made for escalation in fuel prices likely over the 5 year operating period, or if insulation is likely to last until the renewal of the boiler, and sufficient pipe insulation might reduce the size and hence the capital cost of a replacement plant, serious consideration should be given to 90 mm insulation which is almost the same total cost. The protection of pipes connecting widely separated buildings has received considerable attention in recent years with the development of district heating. The traditional practice of supporting pipework in covered ductways is expensive and has not always proved to be satisfactory. If the duct becomes flooded, even slight gaps in the waterproof covering of the insulation, will permit the insulation to become soaked and the covering itself may then serve to retain the water, leading to rapid external corrosion of the pipe. Failure to seal the covering material adequately at flanges or at pipe supports is a common cause of this defect. For this reason, it has become common practice to lay the pipe or pipes directly in the ground heavily insulated, rendered corrosion proof by means of a waterproof case. Frequently valves and controls in pipe lines are ignored and present sources of considerable heat loss. Valve boxes are well worthwhile, and on high temperature pipe lines "heat bridges" formed by supports, etc., might receive similar attention. The painting of thermal insulation can have an appreciable effect on heat emission. Colours may sometimes be governed by the firms policy on identification, but the use of a paint of low emissivity, such as aluminium, can reduce the quantity of heat, which would otherwise be lost by as much as 16%.

Apart from considerations of thermal economy, insulation may sometimes be necessary to protect persons from injury, due to physical contact with hot surfaces. B.S.C.P. (British Standard Code of Practice) 3005: 1969 recommends that the temperature of a non metallic surface within reach of a permanent working floor should not exceed 65°C (150°F) reduced to 54°C (130°F) for metallic finishes. At high levels, but where access is possible, a limit of 49°C (120°F) is prescribed. It goes without saying that a high standard of maintenance is desirable. In particular a look out should be kept for pipe leakages concealed by insulation. It is possible for rapid corrosion to take place in such situations and for a dangerous situation to develop.

In conclusion other sources of heat loss are summarised below:

1. Badly or loosely fitted insulation or insulation having lost its original bulk density, due to compression.
2. Heat lost from Boiler flue gas surfaces, creating corrosion risks.
3. Heat lost from inadequately insulated warm air heating supply ducts.
4. Heat gained by chilled water pipes or chilled air ductwork, increasing refrigeration plant load and energy consumption.

77

4
The Various Fuels Available

THE VARIOUS FUELS AVAILABLE

Industrial and commercial fuel users are mainly concerned with three basic primary fuels, which can be very briefly summarised as having advantages and disadvantages as follows. Full analyses, viscosities, calorific values and other properties can of course easily be obtained from the suppliers.

1. *Solid Fuels*

    Usually coal, but smaller central heating boiler plants may still use manufactured fuels such as coke. However the almost complete conversion of the gas industry to North Sea natural gas has eliminated the sources of 'gas coke' and compelled users to convert to other fuels, so that users of coke or other manufactured smokeless fuels are now in the domestic or very small commercial boiler field, really outside the scope of this publication.

    Coal is the easiest fuel to store, requiring no tanks or vessels, and little fire precautions; however, as discussed later it can be very expensive to store. With the run-down of certain coalfields, low volatile and dry steam coals have almost disappeared from the larger users' choice and the need to direct prime coking coals to the steel industry means that fairly high volatile bituminous coals are the solid fuels primarily available. Some mechanical stokers, such as the underfeed type, require closely graded coals, without high coking tendencies, to give best results, although others are not so selective. Handling and conveying the coal from point of delivery to the actual furnace can be dusty and rather expensive, and techniques such as pneumatic conveying or fluidised-bed combustion have failed to be developed fully, although full of promise. However, the main problem with solid fuel firing is the ash content and the need to ensure some cleaning of the exit gases to remove grit and dust lifted from the firebeds of modern firing systems. Ashes have to be removed from furnaces by hand in smaller plants, or by mechanical means in larger ones, and even though this may be achieved in modern packaged boilers by extractors or by allowing ashes to drop from the grate through the boiler to a container below, the cost of transfer to a bunker or

81

ground level storage and eventual loading into vehicles for disposal is appreciable, and represents one of the main reasons why coal should always be somewhat cheaper per therm of heat content than fuel oils or natural gas.

While there is some variation in quality, involving not only calorific value but also ash content, coking properties, moisture content and grading, great improvements in quality control have been made in recent years and apart from periods of scarcity due to transport or production problems it is fair to say that reasonably consistent fuels are now delivered to users.

2. *Fuel Oils*

Due to the refinery processes carried out on the crudes, quality control is much easier to exercise; there are relatively few refineries compared to collieries and so consistency of calorific value and analysis is taken for granted by users. Storage facilities appear at first to cost more than with solid fuels, but see later comments. Smaller users have become used to distillate fuels requiring no heating in storage or at the burner, but since the end of 1973 these have increased in price, both per therm and relative to the heavier fuel oils.

The heavier oils, more usual for the larger users, require some heat in ready-use storage and further preheating immediately before burning and this involves both capital expenditure and running costs. Delivery however is easier and generally cleaner, and there is little ash nuisance, apart from occasional tank de-sludging. Some fire precautions have to be taken and burners have to allow for flame failure detection and cut-off of fuel and also purging periods (although these are often excessive, as discussed in Section 1.)

All fuels oils contain sulphur, ranging from around 0.5% for gas oil to 3.5% for the heavy grades, and as discussed in Section 2 trouble can occur in hot water boilers, in flues and chimneys if metal wall temperatures are so low that acid condensation can cause corrosion and incidental emission of smuts.

3. *Gas (Mainly natural gas)*

Much of Britain is now supplied with North Sea gas, double the calorific value of 'town gas' made from coal or cracking of oil, and this has become a bulk fuel in recent years. In 1974 as this Section is being written, there is a temporary plateau in in supply and further large industrial loads are difficult to supply until further pipelines bring large quantities from the more recently discovered fields. Even with these further supplies, many industrial plants may have to accept 'interruptible' tariffs, under which gas supplies may be cut off for quite long periods in Winter. Very large sales of gas have been made for domestic, commercial and school heating, resulting in high Winter peak demands and although work is currently proceeding on underground storage etc., this is unlikely to give any appreciable period of supply or overcoming peaks, so that it seems inevitable that many medium and large industrial plants will have to accept 'interruptible' tariffs in the future.

Of course such gas is very clean, is delivered under pressure so requires no further pumping or compressing for most burner systems, has no harmful components such as sulphur, chlorine or vanadium, with efficient combustion should give no solids emission problems, and there is no ash or sludge to be removed. Storage, other than on a national scale, is impracticable, and so plants wishing to guard against interruption of supply must store an alternative fuel and have burners capable of rapid conversion to this alternative.

4. *Other Fuels*

Some users of liquified petroleum gases such as butane and propane are on quite large scale, for example a paper mill and a major metals firm. Storage requires pressure vessels, and larger consumers have to install evaporators to supply latent heat to vaporise the stored fuel. Again this is a clean fuel with only very slight traces of sulphur. Recent price movements have made this an uneconomic fuel for major users, although it remains a useful source of a rich fuel gas for certain processes, and for radiant heaters in fairly small space heating systems.

Electricity is of course a major fuel, although a 'secondary' one derived from combustion of primary fuel or from nuclear power stations. It has advantages in certain processes where controlability and absence of combustion products are of great value, and where expensive materials are processed. However, for bulk steam raising or large space heating systems it is very expensive. However most users regard it as a source of power, via motors, or lighting.

Waste materials with appreciable calorific values can be used to raise steam or for space heating, ranging from very large municipal garbage incinerators, burning up to 50 tons per hour and even supplying steam turbo-generators, to small incinerators burning waste tyres, wood waste or certain contaminated organic liquid wastes, coupled to small waste-heat boilers.

STORAGE OF FUELS

It is argued that the minimum necessary storage is say 1.5 times the normal delivery quantity, so that a new delivery can be accepted while the plant can still be operating on the residual storage. Ignoring domestic and very small office blocks, this probably means a minimum of 5 to 10 tons of solid fuel, since small lorries deliver from 3 to 7 tons, and a minimum of $4m^3$ of distillate liquid fuels, since $2.5m^3$ (500 gallons) is about the smallest economic delivery that distributors prefer to make. With heavy fuel oils the minimum delivery is likely to be $5m^3$ (1000 gallons).

Domestic consumers habitually operate with such small storage quantities since these are still large in terms of time and suppliers' deliveries are usually reliable and prompt. However, for plants above domestic size such storage quantities often represents only a few days of operation. Again if a domestic user runs out of fuel due to forgetfulness in re-ordering or delays in delivery, there is usually only inconvenience rather than loss of money, as the use of portable electric heaters and the immersion heater or cooker for hot water will enable them to survive. Most plant operators however would argue that greater fuel storage is needed than that related to acceptance of delivery quantities; that continuity of process or of heating is important. They will argue that deliveries might be delayed by very bad weather or suspended by industrial action. However storage is expensive, possibly more so than many users realise.

FACTORS AFFECTING COSTS OF STORAGE

The following list briefly considers points affecting storage costs. Some of these may have to be evaluated with the aid of other specialists from the accounting and production departments.

1. The area needed for storage. For example 100 tons of coal stored in a layer 1m deep would require an area of 130m² (156 yd²) plus space to manouvre vehicles when adding or removing coal. Although fuel oils can be stored in quite tall tanks, most sites will require a 'bund' wall round the tank to contain spillage or leakage.

    To keep the height and cost of such walls to reasonable figures, coupled with constructional problems of resisting the thrust if the whole 'bund' is nearly full of oil due to a major tank rupture, it is doubtful whether heights exceeding 3m would normally be used. This means a minimum area of 1m² within the wall for each 2.5m³ of oil (550 gallons) stored, allowing for tank plate thicknesses, pipes, supports, etc. within the 'bund' and a 10 margin in volume.

    This area must have an annual value; it might have been used for other useful storage or building purposes. It may have to be fenced to keep out intruders.

2. The construction work needed. With oil fuels this includes tanks, bund walls, foundations and flooring, any outflow heaters and line tracing, firefighting precautions and lighting. The capital cost of providing these must be converted to an annual charge over the life of the storage system.

    With solid fuels it is of course possible to stock on unprepared ground, even a field, and it could be argued no construction cost is involved. However, serious losses can occur when recovering such stocks, as a 'carpet' of coal usually remains untouched. Also mechanical equipment, such as shovel loaders or grab cranes, will churn up the ground and force coal into it. Losses of 4 to 5% may well occur, and should be 'written off' on initial stockpiling on such sites. It is preferable to lay down a level concrete floor, capable of carrying vehicles, and low retaining walls, which will reduce the tonnage loss on recovery to negligible proportions.

3. The cost of stocking. Extra supervision, use of mechanical loaders to stack solid fuels above lorry-tipping heights, all have to be considered. However the most important financial item is the capital investment in the fuel, which may well have been able to earn interest if invested. Some firms may even be borrowing at high interest.

4. The cost of maintaining stocks. With bituminous coals of fairly high volatile content, the most usual fuel to be stocked industrially, precautions have to be taken against heating and possible spontaneous combustion in the first months of laying down the stock. This is discussed in some length later, but the cost of such precautions must be acknowledged. With distillate oil fuels, maintaining costs are negligible, but with heavy oils heating is necessary, either continuously if all storage is kept in 'ready-use', as for example if one large tank is used, or if storage is in unheated tanks, these have to be heated up immediately before being brought into service.

5. The cost of recovering stocks. Diggers, loaders and lorries, with associated labour, are needed to move solid fuels from stock to the using areas. It is only on recovery that losses into ground, mentioned in item 2, or losses due to crushing, weathering and loss of calorific value can be determined. With fuel oils transfer pumping may be necessary.

6. Possible buying advantages. Some firms may be able to buy fuel at lower prices at certain seasons, and this may justify short-term stocking, such as buying in summer and using in winter. In recent years inflation has also pushed up fuel prices and this must be allowed for when pricing the fuel brought from stock.

One example might be of interest, where coal was stored at a density of one ton/m^2 of overall storage space, and stored for two years:

	Over two years £
Accountant's estimate of land value £4/m^2 year:-	8.00
Concreting area; storage walls; estimate of annual cost of investment/ton storage space:-	1.60
Interest, compound, at 13%, on purchase price of 30 per ton:-	8.31
Estimated extra cost of stocking 40p, and of recovering 60p, per ton:-	1.00
Estimated loss due to crushing and non-recovery 1%:-	
Estimated loss due to drop in calorific value, 1.5%:-	
Estimated loss due to drop in boiler efficiency because of increased fines due to handling and weathering, 2%:-	
Total cost of storage per ton, over two years:-	22.31

The cost of storage is lower than the purchase price of £30 per ton, although much of this is due to the assessed value of the ground area.

Although users of natural gas cannot normally store it, they may well fit dual-fuel plant as insurance, even if on a 'continuous' supply agreement and carry some stock of the second fuel. With an 'interruptible' tariff they must carry appreciable stocks and some firms argue these should be equal to the maximum interruptible period in case this is imposed as one continuous cessation of gas supply and the suppliers of the alternative fuel are so overwhelmed with peak demands on supply, that they cannot top-up stocks as they are being used. Consequently, all the points mentioned must be estimated and added to the tariff cost of gas to obtain the true cost per therm.

Storage of Coal

Freshly mined coal absorbs oxygen when exposed to the air and this is accompanied by the giving out of heat. If conditions are suitable, and if this heat cannot be lost by radiation or by good ventilation the temperature continues to rise and the rate at which oxygen is absorbed also rises, so that the process carries on and the coal gets hotter and hotter. But as the temperature rises the rate at which the heat can be lost also increases so that a balance may be struck, and the temperature steadies, and then as the rate of oxygen absorption slackens with time the danger passes.

If the heat cannot be dissipated however, the temperature continues to rise at an increasing rate until the coal fires. There is a critical temperature in this cycle, and when it is reached, if the conditions are not altered nothing can stop the coal firing. In a heap of coal it is in the range 52 – 80°C, and above this critical temperature, steam and CO_2 and CO are given off and the process of slow combustion starts, gradually raising the temperature further until at about 350°C active combustion starts.

Given access to oxygen, the production of heat varies with the surface exposed, and as for a given weight of coal the surface exposed is greater the finer the size of the particles, then coal containing a large amount of fines is more liable to spontaneous heating than large sized coal. Also one of the main ways of losing heat is by air passing through the stack, and as air can pass more easily through a heap of large coal than through a heap containing fines, which prevent the passage of air, again large coal is not so liable to fire.

In practice it is difficult to ventilate a stack of fine coal adequately enough to prevent overheating, and it is easier to supress ventilation so that the amount of air entering is too small to support any appreciable oxidation. One good way is to pack the heap evenly and blanket it with fine coal. Also if the coal can be kept damp the moisture fills the spaces between the pieces of coal. A wall built round the stack also prevents ventilation, and the stack's long sides should be parallel with prevailing wind to reduce air being blown into stack.

Never put fresh coal on top of old coal that has not been lying for at least 3 months, so that it has passed the active heating stage. If possible metal tubes should be placed in the stack and temperatures taken weekly. If one part rises to say 55°C, it should be watched and temperatures taken more frequently. If temperature still rises and especially if it starts to rise rapidly, delay is useless. The affected portion must be dug out and the hot coal used, or spread very thinly over the rest of the heap. Once high temperatures are reached, spraying with water is not often useful, as the seat of the trouble is probably 1 to 2m deep and the water may channel through one or two easy routes so that the whole stack is not wetted.

If serious heating occurs during the first months of storage, there can be an apprecible loss of calorific value, exceeding 5% However, if the precautions previously mentioned prevent this, the loss rarely exceeds 1% per year, and is greatest in the first year. One other item affecting the efficiency with which the fuel can be used is degradation. The extra handling creates more small or fine coal than in the original deliveries, and the action of weather causes further breakdown. For example, a graded coal, almost all above 20mm size prior to storage, may have 25% below 20mm after two years storage. With many boiler plants this may cause loss of efficiency due to extra riddlings through grate bars and/or poor mechanical stoker performance preventing optimum excess air usage being achieved.

STORAGE OF FUEL OILS

The heavier grades of fuel oil require preheating to raise the temperature to that needed for handling, and further heating for most burners. However, even with light fuel oil installations heating is sometimes applied to cope with very cold wintry weather. The 'cloud point' of some gas oils (35 seconds oils) can be as high as 28°F and indicates the point when some fouling of fine mesh filters can occur. Also although this 'cloud point' is slightly above the 'pour point' an oil might still become so viscous that it will not flow along horizontal lengths of small-diameter feed pipes if the only pressure encouraging it to move is two or three metres of gravity head of oil in the full storage tank, or considerably less when the tank is almost empty. Because cases have been known where light oils have failed to flow from outdoor tanks in extremely cold weather, many installations have incorporated a small immersion heater in the storage tank and lightly-loaded trace heating of outdoor portions of the supply line from the tank to the burner, both usually using electricity in this smaller type of plant.

Unfortunately cases have been seen where the immersion heater particularly is controlled by an air thermostat, to bring it into service when the outdoor temperature is low. Some of these thermostats have been found to be set as high as 45°F (the small heating contractor usually involved with these installations buys 'domestic' type thermostats with minimum settings of around 30°F, and caretakers usually 'set them up a bit' to be safe). In one extreme case it was found that both the immersion heater in the storage tank and the electric tracing cable had been left switched on all winter as the outdoor thermostat was faulty. This load of rather over 2kW had been in use for some 5000 hours, costing the school around £230 per year, and it was impossible to determine how many years ago the fault had first occurred. Fortunately the heaters were fed from the boiler panel so that they were switched off outside the heating season when the whole plant was shut down.

Such systems should be checked, and a good thermostat with low-range setting substituted for any 'domestic' type. Many immersion heaters are of 2 to 3kW capacity; this is quite unnecessary, as a small heater, around 1kW size, situated adjacent to the offtake pipe from the tank, is more than adequate to give the very small rise in temperature needed to keep light fuel oil flowing freely even in the most extreme weather ever likely to be experienced in Britain.

VALUE OF STORAGE

Fuel is stored to prevent plant shut-down, and the portion of Section 1 on the cost of boiler outage should be re-studied to decide what financial loss or loss of amenity would occur if a plant had to shut down for lack of fuel. Unlike sudden breakdowns, fuel can be seen to be running out, so that process plant could be shut down in planned fashion to avoid spoilage of raw materials, and this time of shut-down should represent the very minimum reserve of fuel so that in times of difficult deliveries, the decision can be made and damage to materials or furnaces avoided.

After such a planned shut down, loss of production can be assessed as a sum of money, and compared to the cost of fuel storage. It will usually be found that where the cost of fuel is only a very small percentage of total costs, there is a good case for carrying considerable reserve stocks, since even though the cost per ton from stock may be more than double the cost of new deliveries (as has been demonstrated by the example) the increase in production costs may be negligible. However, some processes involve very high fuel costs and some firms may find that with a correct assessment of the real costs of storage, it may actually pay them to shut down rather than carry large reserves.

Office workers have demonstrated their ability to manage reasonably well without normal heating, and emergency substitutes, such as electric heaters, may be valued at £300/m² in the centre of large towns. It may be argued that fuel shortages may cut electricity supplies too, so that emergency heating could not be available, but very few heating plants can run without electricity, so reserve storage would still be useless in such circumstances.

To sum up, the whole point of the last few pages is to try to force fuel buyers to decide on correct stock levels; uselessly high reserves cost money and act against the economies in the use of fuel; and on oil-fired plants to suggest how to minimise heating costs in storage.

One or two storage tanks have been seen where these immersion heaters have been placed through a boss in the side of the tank well above the off-take pipe, and without any limit thermostat inside the heater casing to shut off the supply if overheating occurs. Such installations are very dangerous, as if the tank is almost empty the heater can be above the oil level and become very, very hot. NIFES know of at least one storage tank fire due to this reason. It can only be urged that any installations like this should be altered instantly.

If, with heavy fuel oil, a single very large storage tank is installed to meet the dual role of immediate user requirements and reserve storage, then the whole contents have to be kept at elevated temperature, say 30°C with some grades, even though an outflow heater may be used to top-up the temperature of outgoing oil to say 45°C for pumping and transmission to users. This may cause unnecessary heat loss compared to having two or more tanks, since only one tank need be heated and in service. If this is done, the reserve tanks should have adequate sludge coil areas, and if warm oil is recirculated from the users by a 'spill-back' or return loop system, valve systems should allow this return to be directed to any tank as soon as it is brought into use. Even with good coil areas, a reserve tank can normally only be heated up at most at 1°C/hr., so that some hours, or even a day, may elapse in cold Winter conditions before a reserve tank of heavy fuel oil can satisfactorily take over service duties. However, loss of deliveries usually occur with sufficient warning to allow for this warming – up period. Another advantage of a relatively small tank for heavy oils is that deliveries are usually of hot oil and if these are put into the 'service' tank and used relatively quickly there is a considerable reduction in storage heating costs as the heat in deliveries may provide most of the input needed to keep the tank at handling temperature.

Some older plants used a 'hot storage' system in which the whole contents of the storage tank were kept at 'handling' temperatures by coils, and outflow heaters not used.

This increases the heat losses appreciably. For example, a 9m long by 3m diameter uninsulated horizontal storage tank kept at 25°C (using an outflow heater to raise the outgoing oil to 45°C) would use around 18kg/h (40lb/h) steam over the 5000 hours of a typical heating season, but if kept at 45°C the consumption would rise to 50kg/h (110lb/h). The extra steam usage just to offset heat losses, with no oil throughput, would use just over 11 tons of additional fuel oil even with an efficient boiler plant, say £1,100 per year at typical 1979 prices Where firms have such systems they should either instal outflow heaters and reduce bulk storage temperature, or consider insulation of the tank. Typical insulation would reduce heat losses by 75%, saving £825 per year, whereas the outflow heater and uninsulated tank at lower storage temperature would give the £1,100 saving mentioned above. While each case should be examined, usually the cost of insulation is much higher than the installation of an outflow heater; however the tank usually has to be drained to allow a larger hole to be made for insertion of the heater, while insulation can usually be applied without interference with plant operation, so that if the plant has just had its annual

92

shutdown, tank insulation may cause savings to accrue for a full heating season before an outflow heater could be installed. Again existing 'bund' wall arrangements may make it impossible to arrange space for withdrawal and servicing of an outflow heater and insulation would then be the only choice.

Systems may include electric heating elements in outflow heaters and electric tracing of flow and return lines between the tanks and burners so that the plant can be started up from cold. While this may be a very useful facility, care should be taken that electricity is not used unnecessarily as its cost per therm of useful heat is about three times that by using steam or hot water from a boiler plant. Surprisingly often plants are inspected where the thermostat controlling electric line tracing is set slightly higher than the outflow heater steam supply thermostat, (or steam line tracing, where this is installed) resulting in almost continuous use of a few kW of line tracing electricity.

Thermostats should either be checked carefully and frequently, or the electric heating thermostat connected via a temperature relay on the steam tracing lines so that only one system can be in use at any time.

Some plants have a 'ring main' system in which surplus oil not required by the oil burner (or burners) is returned to the outflow heater or storage tank. Very often the type of burner installed has electric heater elements to provide the final preheating for correct burner operation. Such heaters are usually of 2 to 8kW capacity per burner, and unexpectedly large amounts of electricity may be used if oil is heated to the minumum handling temperature by the outflow heater. Quite useful cost savings can be made by raising the setting of the steam supply thermostat on the outflow heater, if possible to the actual temperature needed at the burner nozzle. If the outflow heater has insufficient heating surface to achieve this, an in-line heater could be installed in the boiler house.

For example, at one plant with four burners on a ring main, with oil supplied to the burners at 45°C it was found by installing check meters that the average electrical consumption for heating to final temperatures of 80°C was 10kW. This plant operated a process load 90 hours/wk, all year round. By raising the ring main temperature to 80°C, electricity consumption dropped to 1kW (for topping-up occasionally to offset line losses). Electricity at 70p per therm was replaced by steam costing 30p per therm marginal cost (labour etc. remaining constant) and the annual saving was £560.

Where steam is used for tank coils or outflow heaters it is usually recommended that the condensate is not recovered for boiler feed water, to prevent any risk of traces of oil entering the boiler. Only too often this means condensate is rejected direct to drain. At a typical steam pressure of 4 bar, this means 23% of the heat in the inlet steam is lost. A second pipe coil could be fitted in the storage tank and condensate circulated through it to recover some of the heat – with oil stored at 25°C the condensate could be cooled to 35°C and the loss to drain would then be only 5%. The steam consumption for oil heating would be reduced by 20%.

In some cases the condensate could be used via a heat exchanger to preheat incoming cold water for boiler feed or process, and it could then be cooled to below 35°C with rather greater savings but possibly higher capital cost. One or two firms have installed a special hot water tank for such condensate, and when nearly full, visual inspection checks that no oil traces are present before the tank is emptied into the main boiler feed tank. This recovers not only most of the heat, but also the value as 'soft' water.

DEPOSITS DUE TO FUELS

With natural gas, combustion should give clean gases, so any build-up of soot on boiler heating surfaces is due to incomplete combustion and indicates very bad burner performance which should be rectified at once. However with liquid, solid and waste fuels there are usually chemicals (e.g. sulphur and chlorine) which give acid gases and which in turn can cause severe corrosion in cooler parts of the plant, flue and chimney system. Solids may also be present which do not burn to clean gases (e.g. vanadium and ash in heavier grades of oil). Again, with solid fuels, combustion may not be fully completed by the time that solid particles leaving the furnace are quenched by contact with a water-cooled surface so that deposits may contain small amounts of unburned fuel.

The harmful effects are twofold. First all deposits on the gas side of heat transfer surfaces act as insulation layers and reduce plant efficiency. This reduction is progressive and can eventually prevent the unit reaching the required output, mainly due to reduced heat transfer rates, but also with gross layers of deposit due to loss of draught. Return to normal performance requires cleaning of these heating surfaces. Although this can be done by hand, usually at shut-down or 'banked' periods, this can be expensive in labour cost. The periods of operation between manual cleanings can be extended by the use of soot blowers which use steam or compressed air. Steam usage can be high, and where nozzles have been allowed to wear and operating times extended beyond maker's recommendations in the erroneous belief that this increases the cleaning action, over 3% of the total steam output of the plant may be used for soot blowing. Correct sequencing of soot-blowers, usually to move loosened deposits progressively from combustion chamber to flue exit, and timing can be carried out by automatic time controller, but usually this is considered a 'luxury' on small plants. However with really good control the amount of steam used for cleaning can be reduced to below 1% of total output. A saving of 2% of annual fuel bill is worth achieving. Compressed air is usually more expensive in use than steam.

Second, when boilers are shut down, particularly after a heating season, the deposits may be contaminated by corrosive compounds and severe corrosion of metal can occur if the boiler is allowed to stand in a cold, dirty condition. Boiler cleaning should be carried out as soon, and as thoroughly, as possible, preferably as soon as the boiler is cool enough for reasonable comfort of the cleaner! If boilers are to be left for some time, say over the summer, a neutralising spray wash, such as with a 'milk' of slaked lime can be very useful. Although Section 1 argued that running costs were more important than initial boiler cost, there is every justification for trying to extend the useful life of the boiler. Such a spray should also be applied through any available inspection doors to the flues and even inside the chimney. Some oil-fired plants in Britain have injectors which allow a metered trickle of powders such as dolomite to enter the boiler flues and pass through with the gases to neutralise acid gases. While the boiler is running this can give a useful degree of protection to flues and inside the chimney where condensation may occur, but it may not be realised that such an injector can be used very usefully after thorough boiler cleaning, running the burner without oil supply for a minute or so at lowest possible air throughput and with the injector at maximum powder input, to coat the boiler tubes, flue walls etc. with powder to continue to neutralise any corrosive materials.

Soot blowers were often fitted to solid-fuel fired Economic type boilers but manufacturers of more recent 'packaged' boilers using solid or liquid fuels have often omitted these, arguing that higher gas velocities reduce deposition of solids from the flue gases. However increased availability may justify their use on plants where no standby is available, and improvements in efficiency can also repay their cost even where spare plant allows shut-downs for Manual cleaning. However, improvements in efficiency must exceed 1%, averaged over the whole boiler operating period, before the steam usage is paid for, and as shown in Diagram 11 (Section 2) this means the blowers must maintain the average flue gas exit temperature at least 20°C lower than without their use.

If the boiler is operated on normal load pattern after a thorough cleaning of flues and tubes the time for the gases leaving the boiler to rise by 40°C above the 'clean' temperature can be noted. If this is quite rapid, say a single shift, and after the stoker or burner has been checked to ensure that the rapid build-up of deposits is not due to incorrect combustion, then soot blowers should be considered. If it takes more than a week for this 40°C rise, then it may be preferable to use manual cleaning at weekly intervals when the boiler can be shut down for an hour or so. The use of compressed air percussion lances or electrically driven cleaning tools may reduce labour costs on larger plants.

There may be other reasons for preventing deposits which encourage the fitting of soot-blowers. For example, with waste-heat boilers on incinerators the boiler may be serving a dual purpose. In addition to recovering useful heat from the burning of a waste or obnoxious material, the boilers may be reducing the gas temperature to allow installation of an efficient gas cleaner, such as an electrofilter. Unless constructed of very expensive heat resisting steels, or irrigated with water to cool the metal plates (which may give an awkward effluent problem) here is a limiting maximum temperature for good operation and long life, say around 250°C. If fouling of the boiler causes this temperature to be exceeded then the plant throughput may have to be reduced until temperatures drop. Such installations may justify much greater expense in soot-blowing equipment than indicated by the value of the recovered heat. In some waste-burning boiler plants, particularly burning some trade wastes or town refuse, very strongly bonded deposits can occur due to semi-molten slags containing tin, lead and boron being carried in the combustion products and solidifying on water-cooled surfaces and a better remedy than soot blowers may be to reduce the inlet gas temperature to the boiler. This can be done by recycling cool gas from the exit of the gas cleaner, admitting cold air into a mixing chamber just before the boiler, or spraying water into such a chamber. Experience suggests that gases above 750 – 800°C will cause excessive fouling problems and give deposits that are hard to remove with such wastes.

Some recent damage has been reported to fibreglass flues and chimneys even though chances of reaching the normally accepted safe temperatures for such material did not seem to have been likely at design stage. It is felt that excessive deposits may have been permitted in the absence of good instrumentation, and the gas temperatures entering the flues and chimneys may have risen quite considerably above the 'design' figures.

DUAL & TREBLE FUEL PLANTS

Coal, fuel oil and natural gas are so very different in characteristics and combustion behaviour that many compromises have to be adopted if a boiler or furnace is to burn more than one fuel. Dual and even treble firing has certainly been practiced for many years, but usually on large specialised plant, particularly in the steel industry, where full time attention was possible. A first study of the characteristics of the fuels might suggest that in any given furnace configuration they are incompatible.

For example, fuel oil in a normal burner gives a highly emissive flame and heat transfer by radiation varies in proportion to emissivity, and in the actual combustion volume of the boiler (furnace tube of a 'packaged' Economic type boiler) radiation is the major method of heat transfer. Typical emissivity coefficients are:

Pulverised fuel	0.7 – 1.0
Fuel oil	0.25 – 0.6
Natural Gas	0.15 – 0.25

This means that heat transferred in a typical furnace tube may be 25 – 30% less with natural gas than with oil, so that the temperature of the gases may be considerabley higher at the end of the tube, and on the rear tube plate at the entry to the first pass of smoke tubes. Some troubles have occurred due to this, overheating of the rear tube plate causing undue erosion and leakage if the boiler were run at the same maximum load as was possible with oil.

However, in the cooler convection zones, heat transfer by natural gas is in the region of 15 higher than with oil, so that this helps to offset the lower emissivity in the radiation zone.

Also the question of fouling of heating surface is important, as these remain very much cleaner on gas, so that there is less fall off in heat transfer. Typical 'fouling factors' by researchers are:

Natural gas 1.0

Fuel oil 0.9

Pulverised fuel 0.7

referring to proportion of heat transferred in practice compared to the transfer when boiler is absolutely clean.

Thus for a boiler with good balance between radiation and convection heating surface, the exit temperatures to the chimney will be almost indentical, the only real difference being at the exit of the radiation zone, where with gas there might be a considerably increased temperature.

Although much work was done on pulverised-fuel firing of shell boilers, it seems doubtful whether this will be revived in order to obtain dual or treble fuel firing using coal as one fuel. The capital and power costs of pulverising, the fouling of heating surfaces since all the ash passes through the boiler, the need for very high efficiency gas cleaning equipment and great difficulties in preventing general dusty conditions round the plant are all serious drawbacks. Present equipment tends to use a grate system (e.g. 'Vekos', chain or coking stokers) which is either capable of rapid removal so that a gas or oil burner can be fitted, or over which a burner can fire. If, as seems possible, some industrial plants reconsider solid fuel firing, the fluidised bed system may be further developed.

When considering dual fuel applications on large boilers with superheaters, care must be taken to consider maintenance of superheated steam temperatures, since this will usually be passing to a turbine. The steam consumption of this will vary considerably with change in superheat, and indeed considerable elevation in temperature may cause blade damage. If the plant is a water-tube boiler and the superheat is a convection type, situated after the combustion chamber, the use of natural gas to replace fuel oil or coal may well cause such an increase. However if a radiant superheater is installed as part of the combustion chamber wall, or pendant into it, there will be quite a drop, which typically may be $50°C$. With a typical condensing turbine, this may increase steam consumption by approx 8% for the same power output and there may be problems at the low pressure end due to increased steam moisture content causing blade erosion.

With a back pressure set where there is a fixed demand for the exhaust steam for use in process or heating, power output typically will drop by nearly 10%. Thus even though the boiler efficiency may remain the same and the heat input per unit mass of steam may drop by 5%, there is a nett increase of about 4% in fuel consumption for the condensing case, and the need to find 10% of oringinal power needs from other sources in the back pressure case. An ideal superheater for dual fired applications should combine radiation and convection effects so as to give a reasonably constant steam temperature whichever fuel is in use. Advice should be taken before converting water-tube boilers with superheaters to other fuels where power generation is involved.

DISPERSION OF COMBUSTION PRODUCTS

Plants should have chimney heights sized according to the 'Memorandum on Chimney Heights' published by HMSO, and where solids emissions are involved these should be limited to the hourly weights laid down in the 1971 Regulations of the Clean Air Act 1968, or any legislation that may follow the suggested limits in the Report of the Second Working Party on Grit and Dust Emissions, published 1974. Obviously these heights and limits can be applied when designing new plant, but what can be done with existing plant to reduce possible nuisance? The maximum ground level concentration of gases from the chimney, such as SO_2, and to a great extent the maximum deposition of the finer dust, is inversely proportional to the 'effective height' from which the gases from the plant start to disperse. This effective height consists of two components, the actual height of the chimney plus the height to which the gases rise above it, due to exit velocity and buoyancy of the plume if warmer than the outside air. This additional height is governed by mass and temperature, increases in which raise the distance to which the plume rises.

To improve dispersal therefore, any or all of the following steps can be taken:

1. It may be possible by insulation of the flues, fan casings and the stack itself, to increase the temperature of the gases quite considerably, this in turn will give a higher thermal plume and lower ground level concentrations.

2. Any air infiltration after the combustion chamber itself will have a bad effect because it cools the gases and the flues. Particular offenders are gas stabilisers and on/off oil burners which tend to allow quantities of cold air to pass through the system when no fuel is being supplied. An automatic damper with draught controllers is better than a draught stabiliser, and a modulating burner better than an on/off type. Any obvious gaps in flues or around the dampers can be sealed and the sum of all these alterations is to increase the gas temperature. Also spare boilers need positive isolation by gas-tight dampers.

3. If the gases leaving the top of the chimney are at very low velocities, then wind effects around the chimney may cause down-draughts dragging volumes of gas down the lee side of the chimney. Cold air can even pass down the inside of the chimney for a surprising distance. Minimum gas exit velocities, at the lowest likely plant load, should be of the order of 6m/sec. to prevent this. Higher velocities than these can further improve the position. In extreme cases venturi fittings have been used where, by proper design and very high gas velocities, improvement has occurred.

4. In some Industries, such as Chemicals, gases such as SO^2 or chlorine may be given off in small quantity from the process. If these gases were led to another chimney already carrying a large mass of gas, then the thermal plume effect might give a greater rise to the noxious gases than would occur if they passed into their own small stack. Similarly with the boiler house it is better to combine the gases from all the boilers into a single stack than have individual chimneys.

5. If after considering all the previous items the ground level concentrations are higher than desirable, increasing the height of the stack may be the only solution. Even a fairly small increase may have a pronounced effect. Unfortunately such increases are usually very expensive, as the extra weight may involve new foundations or supports for the chimney.

6. Surrounding buildings or sloping land may cause turbulence near the chimney and bring the gas down to ground level very quickly and in much greater concentrations than the Equation would suggest.

7. It is surprising to see new metal chimneys with 'Chinese 'hats', a shallow cone held by spacers a small distance above the top of the chimney. The common argument is that this

is to prevent rain penetrating down the chimney when the plant is idle. Exactly why this is so important is not clear, since a chimney with a sloping plate at the base leading to a drain syphon will not allow water to stand in it and it can be argued that if heating plants are idle over the summer, dampers should be left open and there will usually be a current of air flowing up the chimney by wind aspiration that will tend to dry the inner wall. However these 'hats' act as a baffle while the plant is running, completely destroying upward velocity and by dispersing the solid gas plume largely destroying any thermal lift. Their removal can give considerable improvements in despersion of gases and dust.

5
Tariffs and Charges

1. Introduction

It is now almost 100 years since the first Act of Parliament was passed concerning the use of Electricity in this Country and which was entitled the 'Electric Lighting Act, 1882'.

At that time, coal gas was the energy generally utilised to provide illumination in the home, office, industry and for street lighting.

Today, coal gas is virtually extinct, natural gas from the North Sea and other areas piped across the Country together with manufactured gases such as propane and methane, serve the needs of home and industry for heating purposes generally, and gas is no longer the energy source for illumination except in small pockets of the Country.

Advances in Technology have progressed at such a rate during the past fifty years so that now man is dependent on electricity in order to exist.

However, basic fuels such as coal, oil and gas are required to provide the energy necessary to generate electricity, and the cost of the basic fuels is reflected in the cost of electricity to the consumer. This has been particularly evident with the increases in charges in October 1973, May 1974 and subsequent further increases.

The advent of Nuclear energy for the generation of electricity brought the hope of cheaper electricity and the commencement of construction of the first nuclear generating station in the early 1950's brought a forecast, by a spokesman of the then British Electricity Authority, of electricity costing less than a ½d (0.21p) per kilowatt hour unit. Whilst this order of cost was perhaps somewhat optimistic there was no reason why the forecast of cheaper electricity could not have been brought to fruition by the 1980's but sadly the dream did not allow the intervention of politics. This intervention in the early 1960's led to the halting of the programme of additional nuclear stations and also restrictions of research in the field of atomic energy. There quickly followed orders for a contraction of the coal industry resulting in a shortage of the fuel which became more serious as time went on and was further aggravated by unrest in the industry caused by the contraction.

Prior to the state of emergency in late 1973 due to the shortage of basic fuels which resulted in a three day working week for non-essential industry, recognition was given to the need for re-assessment of a programme for nuclear station construction. Disagreement in both political circles and the electrical supply industry regarding the type of nuclear reactor to be used resulted in serious delays in forming a new construction programme. Despite the introduction of a Secretary of State for Energy, a decision on the U.K's future nuclear reactor policy, promised for January 1974, then before Easter 1974 was further delayed until mid-1974.

Delays in nuclear policy decisions affect consumers when it is considered that generating stations using conventional fuels cost more to run than nuclear stations, and as time passes the gap will be ever-widening.

Whilst the capital costs per kW of nuclear stations are between 66% and 200% higher than conventional stations the benefits of lower running costs and self-reliance in respect of prime energy are tremendous.

Nevertheless, due to inflation and world escalation in cost of basic materials such as copper, insulating material and others used in the electrical supply and distribution field, the days of cheap electricity are gone for ever.

Gas is utilised generally for space heating in domestic and commercial premises and is also used for process heating in industry. Natural gas, since its introduction has been fairly cheap to industrial users but unless other major fields are located supplies are somewhat limited and this fact plus the need to maintain the energy balance means that the price of gas has risen rapidly to industry and commence from the mid-70s. Therefore , it is necessary in the National interest to conserve energy, both electrical and gas, by efficient utilisaltion of the supplies available. In the consumer's interest it becomes necessary not only to conserve energy but also to exercise close financial control of the purchase of these energies.

The utilisation of these two energy sources are discussed in later sections; this section endeavours to explain the various tariffs on which charges are based and shows how to evaluate costs to indicate where savings may be achieved.

ABBREVIATIONS AND DEFINITIONS

CEGB   Central Electricity Generating Board

M.D.   Maximum Demand expressed in kW or kVA

kW   Kilowatt – a measurement of useful power and is a product of voltage, current and Power Factor.

kVA   Kilovolt – ampere – a measurement of apparent power and is a product of voltage and current

kVAr   Reactive power – a measurement of the magnetising component of inductive loads, really being useless power

P.F.   Power Factor – the ratio of useful power to apparent power

kWh   Kilowatt hour – unit of electricity consumption i.e. 1 kWh unit = kW consumed for 1 hour

kVArh   Kilowatt-ampere – reactive hour – unit of reactive power consumption per hour.

ELECTRICITY

2. Tariff Structure

Prior to the Electricity Act of 1947 there were some 560 Authorised Undertakings supplying Electricity in England, Wales and Scotland, the majority of whom were local authorities. Thus the cost of purchasing electricity varied widely over the country and accrued profits went into the coffers of the local authorities to provide assistance to property rating funds and because of this, maintenance and standardisation of distribution systems tended to be neglected.

The Nationalisation Act, as it became known, rationalised the supply system and also the administration in Great Britain.

The supply system is now administered by fourteen area boards, twelve in England and Wales and two in Scotland, from whom most consumers purchase their electrical energy.

Each of the area boards in England and Wales purchase electricity from the Central Electricity Generating Board whilst the two boards on Scotland normally generate their own, although these two boards are linked to each other's network and to the network of the CEGB. The interchange of energy between the CEGB and the Scottish

107

boards together with interchange with France, generally provides for security of supplies throughout the whole of Great Britain. Surpluses from privately owned generation plants are imported by area boards but this source of supply is very small.

Each of the area boards and the CEGB are autonomous and are responsible only to Parliament, but they liaise closely with each other through the Electricity Council, and through the office of the Secretary of State in the case of the two Scottish boards.

The Electricity Council came into being with the Electricity Act 1957, and its function is:-

(a) to advise the Minister on questions affecting the electricity supply industry and related matters and

(b) to promote and assist the maintenance and development by Electricity Boards in England and Wales of an efficient, co-ordinated and economical system of electricity supply.

There is a Consultative Council associated with each of the area boards in Great Britain and each Council consists of representatives of local authorities and representatives of agriculture, commerce, industry, labour and consumers in general.

Within the framework of the Electricity Acts of 1947 and 1957 is the section relating to the prices to be charged for electricity.

In England and Wales, the CEGB charge for the supply of electricity to the Area Boards at prices which are in accordance with such tariffs as may be fixed from time to time by the CEGB after consultation with the **Electricity Council.** In turn, each Area Board charge for electricity supplied by them, to consumers, at prices which are in accordance with such tariffs as may be fixed by them after consultation with the **Consultative Council,** established for their area, and the **Electricity Council.**

In the case of Scotland, the two Area Boards charge for the supply of electricity at prices which are in accordance with such tariffs fixed by them after consultation with their **Consultative Council.**

The Consultative Councils therefore are watchdogs of consumers, and their responsibilities are wide and varied, not least of all being to ensure that the cost of purchasing electricity is fair and reasonable and also to act as arbiter in any disputes, in respect of tariffs, between a consumer and their Area Board.

The charges of supplying electricity to Area Boards by the CEGB are based on a Bulk Supply Tariff, the components of which will not vary, except by major tariff revision, but the price of each component may vary from year to year dependent on generating costs, and will be as published by the CEGB.

The Bulk Supply Tariff is made up of the following components:-

1. **Peaking Capacity Charge** – being a cost per kW for every kW by which the Area Board's peak demand exceeds its basic demand.

2. **Basic Capacity Charge** – being a cost per kW of the Area Board's basic demand.

   For the purpose of the above capacity charges the following definitions apply:-

   (a) **'peak demand'** means half the sum of twice the number of kilowatt-hours supplied to the Area Board during the half of maximum system demand before a defined time and the half hour of maximum system after that time during certain periods of the year.

   (b) **'system demand'** means twice the number of kilowatt-hours sent out from the Generating Board's power stations and purchased by the Generating Board from other sources of supply within a single half-hour in the year.

   (c) **'potential peak warning periods'** means those periods for which the Generating Board has on the previous day issued a warning that the maximum system demand may occur. The aggregate of such periods shall not exceed a set number of hours in the year, and such periods shall be confined to be within the hours specified for the Peak Period Running Rate.

   (b) **'basic demand'** means twice the average number of kilowatt-hours supplied to the Area Board in each half hour when the system demand during the year was within ± 1 per cent of 90 per cent of

109

the average of the two maximum system demands specified in (a) above.

3. **Peak Running Rate** – being a cost per kilowatt hour supplied between defined hours in the period November to February of specified dates of the two months, both dates being inclusive, but excluding Saturdays, Sundays, Christmas Day, Boxing Day and New Year's Day.

4. **Day Running Rate** – being a cost per kilowat hour supplied between defined hours on all days throughout the year excluding those kilowatt hours charged at the above Peak Running Rate.

5. **Night Running Rate** – being a cost per kilowatt hour supplied between defined hours on all nights throughout the year.

6. **Fuel Cost Adjustment** –means that the above Running Rates are increased or reduced by an amount determined by the CEGB depending on whether the national fuel cost per tonne in the year is above or below a standard price per tonne.

   (a) **'National fuel cost per tonne'** is determined as:-
   The total cost, inclusive of supply and delivery, of the fuel consumed at all the Generating Board's stations in the period divided by the total number of tonnes of fuel so consumed, and the cost per tonne so obtained shall be multiplied by 26 000 and divided by the average gross thermal value of the fuel so consumed expressed in kilojoules per kilogramme.

   (b) **'Fuel'** means coal, coke, oil and gaseous fuels but does not include nuclear fuels.

The tariffs on which charges are based for the supply of electricity to consumers by the Area Boards vary from Board to Board and, because of the complexity of types of consumers, are numerous, although in the 70's the number of tariffs have been reduced.

However the basic component of the tariffs are similar in each Board area but because each Area Board is autonomous there are slight variations in some Boards of the categories of consumers and there are also variations of tariffs offered to those categories. This section cannot hope to explain in detail all the tariffs offered by all the

Area Boards. It is necessary for the Consumer to study the published tariffs for the area in which he is situated. Broadly speaking then, the tariffs related to the various categories are as follows:-

Domestic

Generally there are two tariffs offered to domestic consumers. With the basic tariff, charges are based on unit consumption per quarter with an initial block of units charged at a high rate and all in excess charged at a reduced rate, or with a standing charge instead of this initial block. The alternative tariff is beneficial to those consumers having some form of electric storage heating. Charges are based on unit consumption per quarter with a low rate for unit consumption during a defined night time period and a higher rate for all units consumed outside that period. In addition there is usually a fixed quarterly charge.

All domestic tariffs now include an additional charge per unit which is related to the cost of fuel.

Commercial

Tariffs offered to commercial users vary according to the electrical load but will generally fall into two categories. The first being a tariff where charges are based on unit consumption, and the second being a two part tariff where charges are based on unit consumption and the maximum demand made on the Electricity Board's system.

The maximum demands may be charged on the highest demand in a month, in a year, or over certain winter months only. All commercial tariffs include an additional charge per unit which is an adjustment for fuel costs.

Industrial

Industrial users are generally offered tariffs similar to those offered to Commercial users and again vary according to electrical load.

However, most Area Boards also offer alternatives related to the seasons of the year and/or time of day. In addition most industrial consumers are entitled to concession in the cost of unit consumption used during defined night time period although some Boards do not publish this

111

facility. A fixed recurring charge is usually made for metering the night unit consumption. It should be noted that some Area Boards impose an increase in cost on daytime unit consumption when allowing concessions for night usage. All tariffs include charges relating to the fuel variation costs.

Agricultural

Tariffs offered to Agricultural users vary from Area Board to Area Board and range from one part flat rate tariffs to two part tariffs comprising charges for assessed demands and unit consumption.

Again all tariffs are subject to additional unit consumption costs related to fuel variation costs.

Power Factor

The CEGB and the Area Boards have to use capital to ensure that there is an efficient supply system capable of containing a given peak load or demand and must also finance the running costs of the system.

Therefore, most tariffs irrespective of category of consumer are framed on a structure having two components one related to capital and the other component related to running costs.

This is clearly shown in the two part tariffs having maximum demand charges and unit consumption charges.

Maximum demands will be measured in kW or kVA dependent on the Area Board.

Where the maximum demands are measured in kVA, charges are made on the highest kVA demand recorded over 30 minutes in any month, quarter or year dependent on the tariff. For a given kW load, kVA load, kVA is inversely proportional to the value of power factor, therefore the charges impose an automatic penalty in respect of power factor.

Where the maximum demands are measured in kW, a power factor penalty clause is usually included in the tariff. With this type of tariff, power factor is calculated from a trigonometrical ratio of kVArh and kWh readings over a given period and is therefore an average value. A penalty is imposed when the average Power Factor is below a certain figure, which will vary between 0.85 and 0.95 dependent on the particular Area Board's determination. The actual power factor figure may not be indicated, however the penalty cost will usually be indicated separately on the consumers account.

From the above it will be seen that good power factor is reflected in the overall cost of electricity and that some consumers are penalised more than others, although it is not obvious with some Area Boards, that penalties are being imposed. There is scope for Area Boards, to have a common policy in respect of power factor and thus remove anomalies which exist for being domiciled in one area or another. Such a common policy would also assist in respect of national energy conservation.

Paying attention to, and improving power factors therefore may well effect savings in the cost of purchasing electricity. How power factors can be improved is dealt with in a later section.

3. **Geographical Variations**

An important factor governing the price at which Area Board's charge for supplying electricity is the revenue return on the capital outlay for their distribution systems. The revenue return for high population density areas is greater than those areas of low population density, thus the proportion of high density to low density areas in relation to the physical area administered by an Area Board will be reflected in the prices charged. Usually areas of high density population are indicative of concentration of industry and/or commerce and therefore those Area Boards having a high degree of such concentrations within their administration area will have a better revenue return than those Boards who have a greater degree of rural areas within their boundaries.

4. **OTHER FACTORS AFFECTING PRICE**

Diagram 16 shows the sudden change in supply pattern that occurred in 1974 when the steady growth in load that had occurred year after year since the early 1950's suddenly and abruptly stopped. After a few years of stagnancy the load has recovered again, but in 1978 had only reached the level of 1974. Such an abrupt change plays havoc with plans for forward building of new power stations and associated employment prospects for firms making large boilers, turbines and associated electrical

113

equipment. Also since power stations take many years to build, the CEGB have extra capacity coming into use without the rising load for which it was ordered. The Diagram shows how, from about 1971, the station load factor has dropped. Any economist will tell us that this must increase the cost per unit when the capital expenditure on these new stations is not able to be recovered by extra load. Tragically the same diagram shows that the

TABLE SHOWING DIVISION OF SALES OF AREA BOARDS

Area Board	% Of Total Units Sold				
	Domestic	Commercial	Industrial	Agricultural	Others
London (LEB)	43.4	36.7	19.0	0.01	0.89
South Eastern (SEE BOARD)	54.1	17.9	22.8	1.5	3.7
Southern (SEB)	47.7	16.3	30.6	2.2	3.2
South Western (SWEB)	48.4	16.5	29.1	3.9	2.1
Eastern	48.6	16.5	30.5	2.3	2.1
East Midlands (EMEB)	35.7	13.4	47.2	2.0	1.7
Midlands (MEB)	36.8	12.4	46.3	1.7	2.8
South Wales	24.1	10.4	63.3	1.6	0.6
Merseyside and North Wales (MANWEB)	29.7	11.0	55.7	1.5	2.1
Yorkshire (YEB)	29.4	9.9	56.7	1.3	2.7
North Eastern (NEEB)	32.5	13.9	50.9	1.6	1.1
North Western (NORWEB)	36.6	13.5	44.9	1.5	3.5
Average % of All units sold	38.9	15.7	41.4	1.8	2.2

growth in load over the years is only partly due to industry – indeed, there can be seen to be a distinct change in the slope of the industrial load line around 1965, and power usage in industry did not increase as much after that year

114

DIAGRAM 16
U.K. ELECTRICITY GENERATION, SUPPLY, ETC. (from 1955)

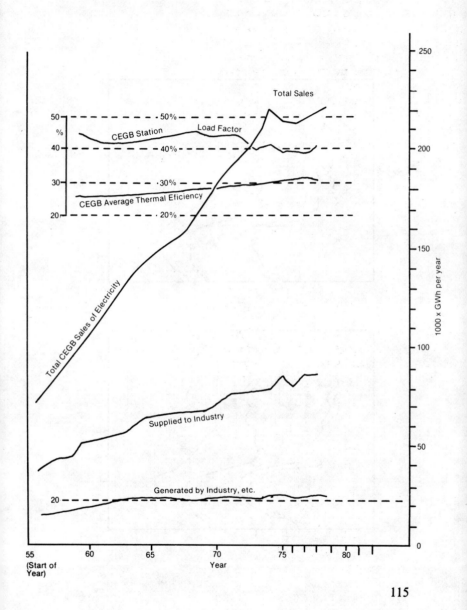

DIAGRAM 17

SUMMER AND WINTER DEMANDS ON CEGB SYSTEM
(Including Minimum and Maximum Demand Days)

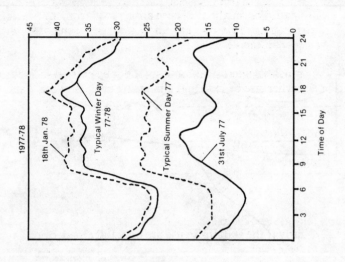

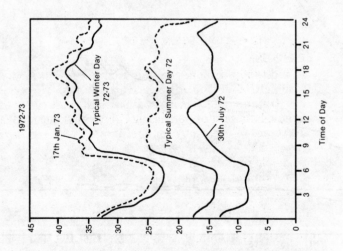

116

as before it.

Diagram 17 shows the pattern of load demand that the CEGB have to satisfy in summer and winter for two periods, 1972/73 and 1977/78 and it can be seen that the pattern for these two years is very similar; hardly any increase in peak demand and hardly any change in night load. This might have been anticipated from the previous Diagram 16 since annual total units generated were almost the same.

20-30 years ago, when industry on single shaft started much earlier (7.30 a.m. or so) the daily pattern showed an earlier rise of load and a morning peak between 7 and 8 a.m. which could be almost as high as the afternoon peak. Changing social patterns have altered this morning pattern and have also filled in the late evening load (due to television, etc.). In turn this has forced Area Boards to alter the night-time period during which off-peak and cheaper units could be offered.

Diagram 17 indicates why maximum demand charges form part of many tariffs, since the capital expenditure to build additional power stations is so high, and particularly if the stations in use are only fully loaded for a few hours each day. Most Boards offer reduced maximum demand charges for the summer period since there is ample capacity then available even allowing for station outage for maintenance.

5. **EVALUATION WITHIN A CONSUMER'S PREMISES**

In most industries, attention is paid to the cost of raw materials, labour, overheads and productivity. Ways and means are constantly being sought to reduce the costs but although the cost of purchasing electricity is included in overheads it is generally considered that little or nothing can be done to reduce this cost.

A simple analysis of electricity accounts however may well indicate that savings can be achieved, sometimes with very little effort.

To carry out such an analysis involves recording essential data from each electricity account received during each financial year. Simple calculations from the data will indicate the average cost per unit and also how efficiently

117

NIFES .... Company ... 'RUBBERSHOES' ....
Area Board ... SOUWEB .... Ref: ....
Tariff ... Monthly Seasonal S2 .... Date: ....
M.D. Charged in kW Service Capacity-600kVA .... Checked: ....

| PERIOD | M.D. | | Units (kWh 000's) | | | Monthly L.F. % | Payments (£) | | | | | | |
Year 1973/74 Month	Restricted Period (1)	Unrest-ricted Period (2)	Normal (3)	O.P. Night (4)	Total (5)	(6)	M.D. (7)	Service Charge (8)	P.F. Penalty (9)	Units (10)	Fuel Variation (11)	Total Energy Payments (12)	Av. costs/ kWh unit p. (13)
APRIL	520		124,100	—	124,100	33	151.00	—	—	439.31	147.47	737.78	0.58
MAY	520		159,140	—	159,140	41	151.00	—	—	547.93	209.40	908.33	0.57
JUNE	490		102,370	—	102,370	29	143.20	—	—	368.80	139.40	651.40	0.64
JULY	500		111,130	—	111,130	30	146.00	—	—	397.00	155.86	698.86	0.63
AUGUST	490		111,432	—	111,432	31	143.20	—	—	396.00	167.65	707.74	0.64
SEPT.	500		138,628	—	138,628	39	146.00	—	—	482.25	208.57	836.82	0.60
OCT.	500		144,036	—	144,036	39	131.00	78.00	—	547.34	167.96	924.30	0.64
NOV.	520		156,940	—	156,940	42	998.60	78.00	—	596.37	289.66	1,962.63	1.25
DEC.	500		95,540	—	95,540	26	961.00	78.00	—	363.05	168.10	1,570.15	1.64
JAN.	460		95,720	—	95,720	28	884.60	78.00	—	363.74	178.18	1,504.52	1.57
FEB.	460		76,640	—	76,640	25	884.60	78.00	—	291.23	160.25	1,414.08	1.85
MAR.	460		93,050	—	93,050	27	884.60	78.00	—	353.59	320.32	1,636.51	1.76
TOTALS					1,408,726		5,624.80	468.00	—	5,147.50	2,312.82	13,553.12	

$$\text{Annual load factor:-} \frac{\text{total kWh}}{\text{MD} \times \text{annual hrs}} = \frac{1408726}{520 \times 8760} = 31\%$$

Av. cost/kWh unit 0.96p
Monthly MD min:- 460 max:- 520

Example 1

118

the consumer uses the electricity purchased. An analysis sheet suitable for recording the essential data and the derived efficiencies and average costs is shown in Examples 1, 2 and 3. The particular figures shown refer to an actual 1974 case, and are not intended to show present-day costs.

The electricity account will indicate the maximum demand recorded in the month, the total units consumed and if applicable, units consumed during 'off peak' or 'night' periods. These figures should be inserted in the appropriate columns 1, 2, 3, 4, and 5. The monthly load factor can then be determined from:-

$$\frac{\text{Total monthly units (5)}}{\text{Recorded MD (1) x hours in month}}$$

The remaining data relating to payments should then be inserted in the appropriate colums. The average cost per unit is derived from:-

$$\frac{\text{Total Energy Payments (12)}}{\text{Total Units (5)}}$$

In terminology for maximum demand periods, columns 1 and 2 will vary from Area Board to Area Board but generally column 1 will be demands in peak periods of day and/or months of the year depending on the tariff.

In order to determine annual costs it is essential that the analysis relates to the Electricity Board's accounting year which may not necessarily coincide with the consumer's financial year. This is particularly so where annual tariffs relate to the highest M.D. in the accounting period. Let us now consider some typical examples of analysis.

Example 1.

A firm engaged in the manufacture of rubber and P.V.C. soled shoes operates on three shift system, 8 hours per shift, 5 day week.

Electricity is purchased from Souweb with charges based on Tariff S.2. This tariff is a maximum demand seasonal tariff where the demand charges are related to the demands in the month and are higher in the peak period winter months of November to March. Unit consumption

119

NIFES
Area Board ...... ROCKWEB ....
Tariff...... Annual M.D./Peak Months
M.D. ....

Company 'NATCHEM'
....

Ref ....
Date ....
Checked ....

PERIOD Year 1973/73 Month	M.D. On Peak (1)	M.D. 24 hrs. (2)	kWh Units Normal (3)	O.P. Night (4)	Total (5)	Monthly L.F. % (6)	M.D. (7)	Service Charge (8)	P.F. Penalty (9)	Units (10)	Fuel Variation (11)	Total Energy Payments (12)	Av. cost/ kWh unit p. (13)
APRIL	900	900	385,140		385,140	69	974.04		8.00	1,617.59	508.73	3,108.69	0.81
MAY	860	860	460,260		460,260	72	974.04		14.40	1,938.09	668.99	3,590.85	0.78
JUNE	880	880	448,420		448,420	71	974.05		-	1,883.36	558.01	3,415.75	0.76
JULY	900	900	401,900		401,900	60	974.04		-	1,687.93	541.12	3,203.47	0.80
AUGUST	860	860	330,300		330,300	52	974.04		-	1,387.26	397.55	2,759.18	0.84
SEPT.	940	940	457,200		457,200	68	974.04		-	1,851.95	533.90	3,360.28	0.73
OCT.	980	-	524,460		524,460	72	974.04		-	1,971.97	692.76	3,639.10	0.69
NOV.	980	-	485,200		485,200	69	1,052.70		-	1,834.91	640.90	3,528.85	0.73
DEC.	980	-	455,110		455,110	62	983.00		-	1,711.00	601.00	3,296.00	0.72
JAN.	1,040	980	557,280		557,280	72	1,573.88		-	2,158.74	736.11	4,469.06	0.80
FEB.	1,040	1,020	525,740		525,740	75	1,042.88		1.60	1,963.90	694.43	3,703.18	0.70
MAR.	1,020	1,600	564,820		564,820	74	1,042.88		12.80	1,937.34	746.07	3,730.41	0.66
TOTALS					5,595,830		12,514.50		36.80	21,939.30	7,319.57	4,184.13	

Payments. (£)

'Annual load factor:- $\dfrac{\text{total kWh}}{\text{MD} \times \text{annual hrs}}$ = $\dfrac{5595830}{1040 \times 8760}$ = 61% Av. cost/kWh unit 0.75p

Monthly MS min:- max:-
on peak 860 on peak 1040
24 hour 860 24 hour 1020

Total payments include 33p/month for hire of 'on peak'

Example 2

120

is charged at a single rate for all units, except that a lower rate is available for units consumed during the night time between 23.00 hours and 07.00 hours.

The analysis for the year 1973/74 is shown in the Table for 'Rubbershoes' and the effect of the general revision to industrial tariffs is clearly shown with the introduction of service charges in October 1973.

Also clearly seen is the steep rise in fuel variation costs. In December with a consumption of 95 540 kWh units the fuel variation costs were £168.10 or 0.18p/unit. By March 1974, these costs had increased to £320.32 for a consumption of 93 050 units or 0.34/unit, almost 100% rise.

The annual load factor of 31% is somewhat distorted due to the fuel emergency which brought about restricted use of electricity, nevertheless the monthly load factors show that there is room for improvement in the use of electricity. Investigations into production methods may well indicate that by some alterations to production runs, the maximum demands could be reduced thus improving the load factors and reducing the demand charges and overall cost per unit.

Consumption during the night time period is not recorded and therefore all units are charged at 0.38p/unit. Due to the three shift system operated, it can be assumed that night time consumption will be approximately one third of the total. As the tariff offers facilities for night consumption to be charged at 0.26p/unit, an annual saving of approximately £560 could have been achieved. This saving approximates to 4% of the total cost of purchasing electricity.

This example shows two areas where savings may be achieved, one being night consumption and requiring very little effort by use of simple analysis. The other, maximum demand, is rather more complex and requires investigation. The use of recording instruments as used by NIFES will show the times of day when high demands are made and can be used also to determine the machine causing high demands.

Example 2.

This firm manufactures chemicals and operates on a three

121

shift system, 8 hours per shift, 7 days per week.

Electricity is purchased from Rockweb with charges based on Tariff No. 8. This is an annual maximum demand where the demand charges are related to the highest demand in the peak periods of day in the peak months of the year. Unit consumption is charged on a sliding scale dependent on the total consumption and M.D.

The firm take supplies at high voltage from the Board and there is no published alternative tariff. Penalties are imposed in respect of power below 0.925.

The Table for Example 2 shows the analysis for the year 1972/73 and the load factor of 61% is indicative of good utilisation although there is room for improvement as a three shift system of 7 days/week is in operation.

A total penalty of £36.80 was imposed due to poor power factors in four months of the year. Whilst this penalty does not justify the capital cost of installing equipment to improve the power factor to within the required limit, it will be necessary for the maintenance staff to keep a close watch so as to ensure that higher penalties are not incurred.

It is estimated that one third of the total consumption is during the night period and if Rockweb offered a lower rate for such consumption a 4% saving in total cost could have been made.

The maximum demands were fairly steady over the twelve month period and this fact plus consumption during the night time periods could well have been grounds for an unpublished or special tariff to be applied. However, the inflexible policies of this Board usually preclude this.

7. **Example 3.**

This firm is a medium engineering manufacturer and operates on a single 8 hour shift system, 5 days per week. Electricity is purchased from the North Midlands Electricity Board with charges based on the monthly industrial tariff. The demand charges relate to the maximum demand in each month and unit consumption is charged on a sliding scale dependent on the total consumption and M.D.

The tariff applied is one which hides power factor penalties and whilst from the analysis the average power

122

NIFES
Area Board ...... NMEB
Tariff ...... INDUSTRIAL
M.D. ...... kVA ...... Authorised Demand 875 kVA

Company ...... 'NATENG CO.'
Ref ......
Date ......
Checked ......

| PERIOD | M.D. | | | kWh Units | | | Monthly | | | Payments. (£) | | | | |
Year 1972/73 Month	KW (1a)	KVA (1b)	P.F. (2)	Normal (3)	O.P. Night (4)	Total (5)	L.F. % (6)	M.D. (7)	Service Charge (8)	P.F. Penalty (9)	Units (10)	Fuel Variation (11)	Total Energy Payments (12)	Av. cost/ kWh unit p. (13)
APRIL	595	640	0.93	111,830		111,830	24	579.00	18.23		509.94	121.48	1,228.65	1.10
MAY	550	580	0.95	108,250		108,250	25	528.00	18.23		493.62	139.67	1,179.52	1.09
JUNE	550	605	0.91	106,600		106,600	24	549.25	18.23		486.09	115.25	1,168.82	1.10
JULY	530	570	0.93	70,800		70,800	17	519.50	18.23		323.21	84.58	945.52	1.33
AUGUST	575	600	0.96	107,570		107,570	24	545.00	18.23		490.52	111.91	1,165.66	1.08
SEPT.	610	680	0.90	114,210		114,210	23	613.00	18.23		520.79	115.32	1,267.34	1.11
OCT.	720	750	0.96	152,810		152,810	27	672.50	18.23		694.48	177.68	1,562.89	1.02
NOV.	760	850	0.89	170,720		170,720	28	757.50	18.23		777.88	198.51	1,752.12	1.03
DEC.	790	850	0.93	147,740		147,740	23	757.50	18.23		673.69	171.79	1,621.21	1.10
JAN.	810	900	0.90	202,730		202,730	30	800.00	18.23		905.58	235.73	1,959.54	0.97
FEB.	760	810	0.94	165,140		165,140	30	723.50	18.23		750.43	192.02	1,684.18	1.02
MAR.	700	800	0.87	163,650		163,650	27	715.00	18.23		743.21	190.29	1,666.73	1.02
TOTALS				1,468,480		1,468,480		7,759.75			7,369.44	1,854.24	17,202.18	

Annual load factor:- $\frac{\text{total kWh}}{\text{MD} \times \text{annual hrs.}}$ $\frac{1468480}{900 \times 8760}$ 19%

Av. cost/kWh unit 1.17p
Monthly MD Min:- 570 KVA   max:-900 KVA
Average P.F.:- 0.92

If P.F. were 0.98, M.D. charge would be £7283, saving £476.

Note: 1) KW demands and KVA demands are shown on the electricity accounts from this Area Board and therefore P.F. is derived from the two recordings.

Note:2) Demand changes are based KVA

Example 3

123

is 0.92, an improvement to 0.98 would have shown a saving of approximately 2.7% of the total cost of purchasing electricity. This indicates that investigation is necessary to ascertain the exact amount of correction necessary for the improvement.

The annual load factor of 19% is low for the type of industry on a single shift system. An improvement in the load factor, which is likely to effect economies in maximum demands, will require some effort, probably by adjustment of production methods.

An area which may well provide economies in maximum demands as well as unit charges, is lighting within the factory. The roof construction of this factory allows a high degree of natural light and if this is utilised to the full by having a strict control of artificial lighting some savings may well be achieved, particularly as the total lighting load is approximately 15% of the total demands of the factory.

From the above three examples it will be seen that by the use of a simple analysis of electricity accounts some economies become obvious whilst other areas for economy require a more detailed evaluation, and which sometimes requires the knowledge of a specialist in this field.

6. **CHANGING TO AN ALTERNATIVE TARIFF**

    If, after completing an analysis, it is considered that the total cost of purchasing electricity is high, then a comparison should be made with charges based on an alternative tariff.

    It will be necessary to complete another analysis sheet after calculating the charges relative to the alternative tariff.

    The comparison could be made by the appropriate Electricity Board on request or advice can be sought from a specialist consultant such as NIFES.

    A change in tariff can only be effected either at the end of a month or year dependent on the tariff and the particular Electricity Board's conditions should be considered if it is found that a change would be beneficial.

7. SPECIAL TARIFFS

Most Area Boards offer facilities of special tariffs to those consumers who have unusual load characteristics such as:-

(a) High demands

(b) High load factors

(c) Night time consumption

(d) Demands in off-peak periods being higher than in on-peak periods.

Most Area Boards will also require a special tariff or modification of a published tariff to be applied where a consumer uses generation equipment and who requests the Area Board to provide supplies in the event of failure of that equipment. No special tariff or tariff modification is necessary if the consumer does not request the standby facilities or if the generation equipment is used only in the event of failure of the Area Board's supplies. However, some precautions against incorrect synchronisation or reverse power flows may be demanded.

All special tariffs are on the basis of negotiation with the Area Board and the advice of a specialist Consultant such as NIFES may be required to ensure that the terms are the best possible in the Consumers interest.

One feature of most British tariffs is that when savings are made in usage, due to improved efficiency, or self-generation, by a firm, the savings are at the cheap part of the tariff. For example, most maximum demand charges are on a sliding scale. Savings would be encouraged if the kVA saved were at the highest rather than the lowest charge. Similarly, units may be saved, for example by cutting out lighting in bright periods of the day and by cutting down machine running time by better work programming, without particular reduction of the maximum demand. If the firm has a high annual load factor and are on 3 shift working plus some weekend maintenance the units saved may be at the cheapest rate, on the typical tariff and their usage of the first block of units/kVA of M.D. will be unaltered.

This is a 'disincentive' to any conservation campaign since the tariffs produce minimum savings. It might be too much to expect the blocks of units and M.D. to be reversed so that costs might rise with increasing

125

requirements, but this would maximise financial savings and encourage firms to invest capital to produce them. However it is difficult to see why the M.D. and unit charges could not be 'averaged' for different sizes of load, as this would still achieve the Boards' understandable requirement to attract large loads while improving the financial savings obtainable by energy conservation campaigns.

8. **GAS TARIFFS**

As with Electricity, the Gas Supply industry was nationalised in 1947 and twelve area boards were formed to administer the industry in place of the numerous local authorities who were previously responsible. Ten area boards cover England, one Wales and one Scotland.

Each of the Area Boards were autonomous and were responsible for the manufacture and distribution of gas. However the introduction of natural gas created problems in distribution by the Area Boards. Natural gas became the prerogative of the Gas Council and the industry had to adapt to the new concept and the new dimensions which had been added.

In 1973, the British Gas Corporation was formed to be responsible for the whole supply industry of England, Scotland and Wales. The Area Boards, in name, and their autonomy disappeared, but for practical reasons administration remains on a regional basis.

Because of the changing situation with the introduction of natural gas, charges to industrial and commercial consumers vary from region to region and indeed even between consumers within the same region, and are charged on a bulk or interruptible supply tariff related to load factors and peak day demand. Generally, the tariff applied is negotiated with the regional authority and is somewhat of a 'one-off' affair.

The marketing practice of the British Gas Corporation is such that each region is allocated a quota of tariff gas to sell each month and this together with the limitations on the available natural gas means that in some areas potential consumers may have to wait before a gas supply can be made available to them.

Generally there are no alternative tariffs on which to base comparisons and therefore there is little scope for saving on tariff costs. As the bulk and interruptible supply tariffs are usually arranged on a contract basis for a fixed period, it is essential that the consumer makes sure that the terms are the best possible to suit his needs and the advice of a specialist in this field will prove to be of considerable benefit.

Since its introduction, natural gas has been offered at a very low cost per therm. However recent proposals, by the British Gas Corporation, to revise prices have meant substantial increases to industrial and commercial consumers.

The new contracts have caused a steep rise in non-domestic tariffs and give a greater flexibility in raising contract prices. Where bulk supply contracts were arranged and signed before the rapid escalation in energy costs of the last two years, total increases of up to 100% could be borne by some individual consumers where contracts are coming up for renewal.

Approximately 60% of all gas sold in the country is purchased by industrial and commercial consumers and of this, some 75% is bought under special contracts. Those consumers will now be faced with some hard negotiations in order to keep down the cost of purchasing gas.

For those consumers not on special contracts, the proposals include for a rationalisation of all industrial and commercial tariffs and generally it appears that the same tariffs will be applicable in all parts of the country instead of the regional anomalies which existed.

In order to effect any savings in the total cost of purchasing gas it is important that the consumer uses the energy efficiently by improvement in production methods and, where gas is used as the heating medium for production purposes or space heating, that heat losses are minimised by the use of thermal insulation.

6
Services Within Factories or Buildings

INTRODUCTION

Although plant services such as water, compressed air and electricity are essential to most factories and commercial buildings, there is often insufficent thought given to the long term development of the supplies, or the most efficient methods of generation, distribution and utilisation.

Developments in the utilisation of pneumatics for both controls and mechanical handling have resulted in improved quality, and often different pressures, of compressed air supplies being required. In some instances this has led to the duplication of complete factory systems, one contaminated and one dry and oil free, resulting in twice the standing losses of a single integrated system.

In the case of water and effluent, these have traditionally been low cost services and "once through" systems have been accepted as standard practice. The rapidly increasing cost of water and the charges now made for the disposal of effluent are making the capital cost involved in the conservation of water economically justified.

COMPRESSED AIR

Compressors and Design Pressure

The generation pressure selected for a Central Compressed air distribution system is most important as it affects both initial capital costs and energy consumption per unit of air compressed. The losses from leakage in the distribution system are also directly related up to 7 bar for the distribution service but for most applications pressures of 5 bar are adequate and where possible should be used. Diagram 18 shows the increased electrical consumption required to generate compressed air at different pressures, and it will be seen that 15% more power is required to generate at 7 instead of 5 bar. In addition to showing the air pressure to power ratio the graph also indicates that rotary compressors are slightly less efficient than reciprocating compressors. There are other

131

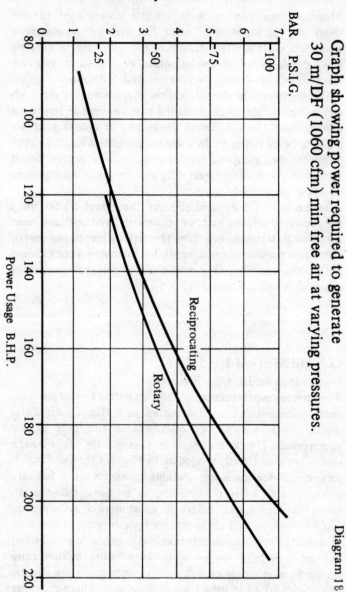

Graph showing power required to generate
30 m/DF (1060 cfm) min free air at varying pressures.

Delivery Pressure

BAR   P.S.I.G.

Power Usage B.H.P.

Reciprocating

Rotary

Diagram 18

considerations which make rotary compressors much more appropriate however for some applications and their overall costs are not necessarily higher.

Diagram 19 shows typical total costs of generating compressed air at different pressures including electrical power, maintenance and labour but excluding capital charges.

If higher pressures are required for a single or very few particular applications consideration should be given to modifying the equipment to use a lower pressure, as this is practical for most pneumatic appliances. Where higher pressures are occasionally essential it is usually preferable to install a small compressor for that particular application rather than upgrade the whole system.

For larger volumes of low pressure air, the most economic solution is to utilise blowers to minimise power consumption. Individual blowers are however preferable to a central distribution system as they avoid the necessity for long runs of large bore pipework and also the wasteful control system which is often employed which maintains the blowers at constant output by releasing surplus air to atmosphere.

If high pressure air is to be used for cooling purposes, effective use can be made of the pressure to entrain secondary air onto the surface being cooled providing some thought is given to nozzle design.

Compressor performance

There are three aspects of compressor performance to be considered:-

(i)   Electricity consumption per unit of output at maximum rating
(ii)   Electricity consumption per unit of output at average demand.
(iii) Capacity of compressor.

The design and manufacturing tolerances obtained determine the basic efficiency of the compressor but the siting can also affect overall performance in respect of inlet air conditions. When commissioned all compressors should be

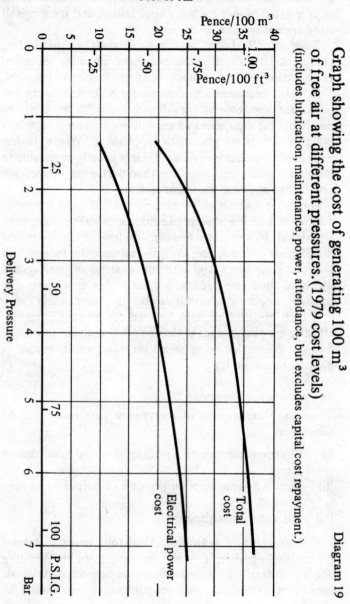

Graph showing the cost of generating 100 m³ of free air at different pressures. (1979 cost levels) (includes lubrication, maintenance, power, attendance, but excludes capital cost repayment.)

Diagram 19

thoroughly checked and records made for future reference. In particular the following should be noted:- Electrical consumption at maximum and no load setting. Cooling water flow rates and temperature gains. Air temperatures and pressures leaving the various stages.

When in operation performance can be reduced in many ways but the most common faults occur due to excessive restrictions across the air intake filters, incorrect setting of the cut off valve on reciprocating machines and general mechanical wear. A log book should therefore by kept showing the primary temperatures, pressures and electrical amps. on full and no-load conditions and inspected regularly to ensure that efficiency and output are being maintained.

Whichever type of automatic control system is used to limit the output from the compressors it will reduce the ratio of air compressed per unit of electrical consumption. Diagram 20 shows the percentage of maximum B.H.P. used for different

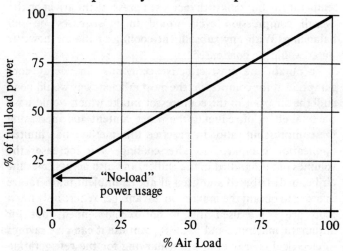

Graph showing the effect of "no-load" power consumption on a partially loaded compressor

Diagram 20

percentages of compressor output. It will be seen that for a compressor loaded to only 30% of its rated output the power consumption will be 42% of its maximum electrical demand. This factor is most significant if a policy of installing individual compressors for specific plant items has been adopted or where more than one pressure is generated at the central compressor plant. In all cases where more than one compressor is used to feed a distribution system, only one compressor should be on control to vary the output, the remainder being set to give maximum output.

Air Quality

In almost all applications clean, dry air should be provided in order to avoid excessive tool maintenance and production delays, or even increased production rejects as in the case of paint spraying.

The inlet air to a compressor system always contains moisture due to the relative humidity of the atmosphere and on a few days each year this can be as high as 85% RH. The location of the compressor house can also affect inlet moisture content if it is sited near to cooling towers or spray ponds. As the moisture carrying capacity of air increases with the rise in temperature, but falls with increased pressure the affect on the overall compression cycle would in most cases be self balancing. With any subsequent cooling of the air however water would be condensed.

To obtain the most effective economies in energy consumption at the compressor the most efficient way would be to chill the air prior to the compressor intake which would have the dual effect of reducing the water content and increasing the compressing ratio. In practice this method has limited application compared to after-cooling as it increases the volumes of air handled in the chiller, and also necessitates the air being distributed and used at above ambient temperature in order to obtain the maximum advantage. Where only hand tools are being used this is not recommended, but for automatic machines and process plant use it can give savings in electrical energy even after allowing for the refrigeration

136

power used.

The normal methods of removing water are therefore with intercoolers between compression stages and aftercoolers before distribution. Water is normally used for cooling and in the past large volumes of towns water have been run to waste with very little temperature gain. Where "once through" systems are used the water quantity should be restricted to obtain the maximum permissible outlet termperature and also isolated when a plant is not in use. Wherever possible closed or evaporative cooling systems should be used, and in large scale applications it is sometimes possible to utilise the heat dissipated for water preheating to process plant.

The use of aftercoolers only ensures dryness to the point where it enters the distribution system and in all installations it is therefore necessary to provide a means of removing water condensed from the air at all positions where the velocity or direction of the air is changed. The alternative is to further chill the air before distribution or remove more water by chemical absorption methods. Where after-chilling is undertaken it should be borne in mind that this has a negative overall effect on the energy cycle and should be limited to applications where the presence of water is very critical indeed. With the correct use of receivers, separators and traps high standards of water free air can be obtained.

As with water the elimination of oil is sometimes critical, e.g. instrument applications and for small pneumatic mechanisms and in these circumstances oil free compressors should be used exclusively. In other applications the presence of a minute quantity of oil can have advantages, e.g. hand tools for lubrication, and for most systems lubricated compressors are quite satisfactory. Excessive quantities of oil in a distribution system can not be tolerated however and strainers and separators should be used for its removal.

Receivers and Air storage
The use of air receivers is essential in most compressed air systems to reduce the effect of fluctuating factory demands on the compressor plant, and also to give a change in velocity of

air in the distribution system so that water, oil and solids can be separated from the air stream and removed. It must be stressed however that air receivers are not analogous to steam accumulators in which the change in state from water to steam at different pressures enables a relatively small volume of water to release a significant volume of steam at a lower pressure. In most compressed air receiver installations the inlet and outlet pressures are substantially the same and it would therefore be necessary to install very large receivers indeed to have any significant effect on a compressor plant with an average demand of say $30m^3$/min (1000 cfm) or more. Where a particular process can operate at a pressure below the central generator pressure and air is required intermittently in relatively small quantities a receiver can be designed to give a constant demand on the compressor plant.

All receivers should be fitted with pressure gauges, an automatic means of removing water, and a drain or access to remove oil and other contaminents.

Distribution System

Due to compressed air being relatively harmless and difficult to see in the form of leaks, losses are allowed to increase to a very much greater degree than with steam or liquid distribution services. In factories using general pneumatic hand tools and automatics, 30-40% of the total air generated will often be dissipated in losses.

For a factory of $50,000m^2$ (540,000 sq. ft.) using air for general purposes only, the total air generated will be approximately $29m^3$/min (1000 cfm) at 6 bar (80 psig) which will indicate losses of $11.5m^3$/min (400 cfm) costing £20,000 per annum on a three shift system.

Although it might be considered that losses of 40% are exaggerated in the above example it should be remembered that this requires the equivalent of only one 3mm dia. hole for each $2500m^2$ (27000 sq. ft.) of working area. In most factories the degree of losses can only be assessed in non-production periods as this is when the sound of escaping air can be heard and the demand on the compressors, without production,

138

calculated or metered. It is surprising how little air metering is done compared with steam.

Diagram 21 shows the losses in cfm for a 3mm diameter hole at different air pressures.

Air loss to atmosphere via a 3.0mm dia. hole at varying pressures.

Diagram 21

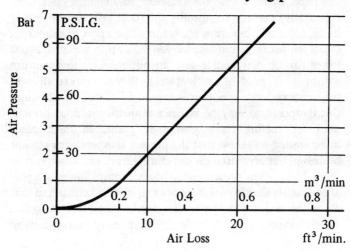

Although regular maintenance is of the utmost importance, the basic design of the system, and particularly the duplication of air services will have a marked effect on the total losses. In most new installations continuous welding of compressed air pipework is recommended. Where water removal is required in a distribution system, automatic valves should always be fitted and checked at regular intervals to ensure that they are not passing. In no circumstances should manual drains be left "cracked" as is so often found the case in some large industrial plants.

COOLING WATER

With the recent significant increases in the cost of water and introduction of effluent disposal charges the economics of

139

water conservation are becoming increasingly attractive.

In the past millions of gallons of water have been used for cooling purposes and, while costs were low, maximum flow rates with minimum temperature gain have been accepted.

Almost all water used for cooling purposes can be recycled either by extracting the heat and reusing in the cooling cycle, or utilising the heat gain for use in another process.

Where water is to be recycled for cooling purposes evaporative coolers are usually the most effective on a cost basis, although they have limitations in respect of minimum outlet temperature and also oxygen pick up. A forced draught tower can be designed to give an outlet water temperature within $4^{\circ}C(7^{\circ}F)$ of the ambient wet bulb temperature which in this country does not normally exceed $17^{\circ}C$ $(63^{\circ}F)$. Although this temperature may be warmer than the water at present used for cooling on a particular plant, in most cases investigation will show that the present temperatures are not by design but are based on the temperature of the water now used. Even when present plant does require lower temperatures, fairly simple modifications can usually be made so that all or the majority of the cooling could take place using the water temperature available from an evaporative cooling tower.

The cost of operating a cooling tower compared to a "once through" cooling system, with a $10^{\circ}C$ temperature gain are:-

Cost per 10 therms of cooling water capacity (assumed hourly requirement)

With Cooling tower:-	Pence
Water make up (6% circulating and evaporation loss)	15
Electrical Power consumption	25
Capital charges	15
Total	55
Buying Towns water at 40p/1000 gallons	223

On a three-shift basis, the annual saving would be approx. £11,000.

PROCESS WATER

Quality Consideration

Usually industrial process waters are derived from either public supply, borehole, river or canal sources of supply. Choice of supply is dependent on availability, quality and economic considerations.

Public supplies of portable waters are becoming increasingly scarce and costly with the result that adaptation of alternative supplies are coming more to the fore. Process water quality is governed by its eventual use. All waters contain impurities in the form of suspended matter, dissolved gases and minerals and unless these are treated they can render the water unfit for use. Broadly, process water is used as boiler feed, cooling water, wash water and as a product constituent. Boiler feedwater make-up quality is dependent on steam pressure and the amount of condensate returned. The greater the boiler pressure the higher is the quality of water required in respect of hardness dissolved solids and alkalinity levels. As make-up water quality affects these, improvements in condensate return, which is virtually distilled water, will minimise the amount and degree of treatment and size of plant required. Dependent on the amount of initial treatment of make-up the cost of not returning condensate including fuel cost can be 225p to 249p per 1000 gallons. Coupled with this, if the alkalinity and TDS level of the feed is high, recourse may have to be made to continuous blowdown in order to maintain the boiler TDS and alkalinity at the desired level. Roughly for every 1% of blowdown a boiler exporting 20,000 lb/hr. at 250 psig (9000 kg/hr at 17 bar) the loss in fuel cost alone can be of the order of £1,000 per annum. If alkalinity in the feed is high, dealkalisation should be considered, as an economic case can often be made as a result of fuel and water costs savings that would be incurred by endeavouring to control the alkalinity level in the boiler water by means of continuous blowdown. Dissolved oxygen and carbon dioxide are undesirable in boiler feed as they encourage corrosion. The solubility of these gases in water is governed by the temperature of the water; the

141

higher the temperature the lower the solubility (i.e. 11 ppm 50° F and 0.1 ppm @ 212° F).

Hence the higher the temperature the better, and in this respect mechanical deaeration should always be considered or alternatively supplementary heating at the feed tank preferably under thermostatic control. Deaeration can be carried out by chemical treatment with hydrazine or sodium sulphite but it is worth noting that to remove 1 ppm of oxygen, 8 ppm of sodium sulphite are required whereas only an equal weight of hydrazine is required. The toxic properties of hydrazine usually prohibit its use especially in the food industry. If sodium sulphite is adopted where initial temperatures are low, an economic case for deaeration plant or feed heating can be made as a result of the cost of chemicals and fuel saved, the latter accruing from the reduction in continuous blowdown that would have resulted with sulphite closing. In the final event, if continuous blowdown has to be adopted heat recovery should be considered as in economic terms it is usually viable. Cooling water quality considerations hinge principally on scale formation and corrosion. Examination of the use could reveal that economies can be affected by increasing temperature rises beyond the typical figures of 5 to 11° C (10° to 20° F). Where these can be increased without undue loss of cooling heat transfer to the user, then flows can be reduced with commensurate saving on electrical power and on water costs. Sometimes higher velocity, with recirculation, will allow a greater temperature rise. Selection and design of materials of pipework and heat exchanger can in some cases minimise the degree of water treatment required to offset scale and corrosion. Careful selection of materials of construction from the corrosion aspect can not be over-emphasised. Should problems still arise, chemical inhibitors such as sodium silicate can sometimes help to curtail or minimise the corrosion. Where hard waters are used as make up in recirculating systems it is normal practice to apply blowdown to control the concentration of bicarbonate hardness at levels around 200 ppm. Examinations of the alternative supplies available could often prove that the blowdown loss could be reduced by changing the supply to a water having a lower hardness. Alternatively if the supply cannot be changed base exchange softening of the make up could turn out to be economically viable.

For low hardness waters chemical conditioning can be adopted. Wash waters should be free from turbidity and organic matter especially in the case of food factories. Clearly choice of supply is important, towns water being usually the better choice. Where water is used as a product constituent similar considerations apply. Economic considerations should take account of what treatment will be necessary to give the required quality i.e. whether coagulation, filtration, dichlorination, softening, dealkalising and other forms of treatment may have to be adopted, and the savings that would result in terms of fuel, supply or extraction costs.

Re-use in other applications

The obvious example of re-using water is by replacing a "once-through" cooling system by a recirculating system using cooling towers. If process demands are high and water resources low, the use of cooling towers will be inevitable. Natural growth of the firm will encourage increased demand which can be satisifed by the installation of small cooling towers to serve individual processes as a short term expedient prior to enlarging any central plant to restrict the spread of "once through" users. The cost of water will make these economic. In cooling processes where heat exchangers are used a review of the final temperatures required can sometimes indicate that these can be connected in series on the cooling water circuit with resulting economies in water usage and pumping electrical consumption. As an aside, the selection of cooling tower size and return can affect performance where waters are collected and pumped, as can be the case where there is a multiplicity of uses the pump should be in constant running with a control valve modulating the flow as opposed to on/off pump control on a float level switch. This will even out and minimise the load on the cooling tower and should be considered when selecting an initial tower size.

Contra-flow washing processes are an application where water can be used resulting in savings of possibly 60% to 70% of consumption. A simple case is illustrated in Diagram 22 for a continuous process of sheet material which passes through bath 'A' in which a chemical concentration of 700 ppm exists. After bath 'A' the material passes through baths 'B', 'C' and 'D' in series, and to each tank is added 500 gallons per hour of make-up water to maintain the solids concentration as shown. The Diagram then shows baths 'B', 'C' and 'D' connected together, water inlet being at 'D' and outlet at 'B'. A flow of 520 gallons per hour in total was found to hold the desired concentrations, giving the same concentration in the last bath with an overall reduction in total flow of 65%. This not only is a worthwhile water saving but reduces the disposal problems considerably.

Effluent Problems

Disposal of industrial effluents is becoming an increasing problem as the river authorities tighten the restrictions in respect of quality of effluents discharged to rivers and sewers and the costs levied for handling discharges increases. Generally it is mandatory that all discharges to rivers are to a quality below or comparable with Royal Commission Standard. Discharges to towns' sewers are generally limited in respect of suspended solids, sulphates, toxic substances such as cyanide, and pH levels. To ensure that the discharge meets the required standard some form of treatment will be required in the form of coagulation and settlement to control suspended solids, pH correction to neutralise any acids and precipitate any toxic metals, and if cyanides are present, conditioning with sodium hypochlorate or chlorine. Where primary settlement is practised sludge dewatering equipment may be necessary together with a disposal source or means of incineration.

Before a decision is made to install effluent treatment plant a careful appraisal should be made of the economies of discharge to available outlets, account being taken of capital,

144

DIAGRAM 22

ORIGINAL PLANT

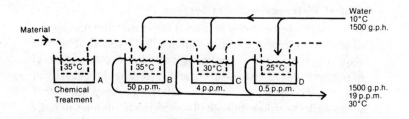

AS MODIFIED

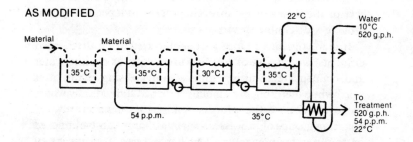

SAVINGS IN A WASHING PROCESS

Hot effluents going to drain can again pass through heat exchangers to preheat process water. Usually this has to be combined with a storage system as the hot water demands usually do not coincide with effluent discharges. Sometimes of course the 'effluent' is sufficiently pure to be re-used in another process.

By modification shown, roughly 1000 g.p.h. of purchased water is saved at say 30p/1000g, or approximately £2000 per year.

Steam consumption reduced to approximately 40% of original, saving approximately £8000 per year.

Treatment plant cheaper as only approximately one-third the volume has to be treated.

chemical, labour, maintenance, electrical, sludge disposal, amortization costs and rates, together with the costs levied by each authority for the discharge facility. The standard of the final effluent may be acceptable as process or cooling water, in which case it would be more economical to recycle the majority of the water for re-use.

Cost of Water and Effluent Services

Before water services can be costed accurately the magnitude of use at optimum conditions should be assessed. This will require getting all processes using water to the optimum level of performance, then measuring and recording the quantity of water used. Water meters are relatively cheap and it is worthwhile installing individual meters in the supplies to large users. Routine records can then be kept to relate consumption to output. Once these relationships are known then regular checking of the records will serve to highlight any wastage. The metered consumptions can be used as a system to evaluate the cost of the water usage of each process independently and provide a discipline to process users

It is worthwhile adopting a similar system on discharged effluents. In the case of recycled users such as cooling water and condensate resulting from steam utilising plant, it is worthwhile metering the amount returned individually where the utilisation of the process is high or, alternatively, by metering groups of processes. Again a charge can be imposed on the user for non-return and hence serve as a means of control.

The costs of water and effluent treatment services comprise:-
(i)   Initial cost of supply
(ii)   Cost of treatment and conditioning which could include any or all of the following:-
    (a)   Primary treatment with coarse filters.
    (b)   Coagulation, settlement, sludge treatment and disposal of any sludge resulting.
    (c)   Secondary filtration (sand or precoat pressure methods)

(d) Based exchange softening, dealkalising or de-mineralisation and treatment, if necessary, of any effluents resulting.

(e) Cost of any waste water resulting from any of the aforementioned.

(f) Cost of any conditioning or inhibiting chemical used such as sodium sulphite, amines, and sodium silicate etc.

(g) Cost of any Effluent conditioning, neutralising, coagulation settlement, treatment, sludge con-ditioning and disposal.

(iii) fuel cost attributable to condensate in the case of non return.

(iv) Electrical power costs in respect of pump and fan power.

(v) Labour including pension holiday pay, overtime, NHS contributions, Bonuses etc.

(vi) Maintenance, materials and labour.

(vii) Rates

(viii) Amortization

(ix) Supervisory, project and design engineering services

(x) Any general or administrative overhead charges

(xi) Cost of effluent discharge facilities.

In establishing some of these costs it may be necessary to make estimates, for example, with pump and fan electrical power but these should be of sufficient accuracy to reflect the true cost of providing the utility.

Once the overall annual cost of each utility has been assessed, knowing the optimum works usage with minimum wastage it should be a relatively simple matter to express the cost to the user, as a price per 1000 gallons supplied or, in the case of effluent, treated.

DOMESTIC HOT WATER

There are many alternative methods used to provide hot water and each type of system has applications for which it is the most suitable. The five main systems used are:-

(a) Separate boilers and systems providing space heating and D.H.W. service.

(b) Common boiler plant service space heating and D.H.W. storage calorifier.

(c) Central electrical or gas heated storage cylinders serving a D.H.W. distribution system.

(d) Local electrical or gas heated storage cylinders.

(e) Local electrical or gas heated non storage system.

Where central system are installed there are also three alternate methods of distribution:-

(i) Dead end system where the tap must run off the cold water in the pipe before hot water is obtained.

(ii) Gravity circulating systems

(iii) Pump assisted systems using small bore pipe returning a small quantity of hot water back to the boiler or calorifier.

For factories where steam or H.P.H.W. is distributed throughout the year for process requirements the provision of a steam/H.P.H.W. heated storage calorifier local to the point of usage is invariably the most economical. Where steam/ H.P.H.W. is only used for heating there is no case for keeping the distribution system live in summer to serve the D.H.W. calorifiers due to the high distribution losses and low thermal efficiency of a lightly loaded boiler. In these circumstances it is normally best to install electric immersion heaters in the storage calorifiers using the steam/H.P.H.W. service in winter and electricity in summer. This system usually has the advantage that the electricity costs are not increased by adding the D.H.W. load to the factories annual Electrical Maximum Demand. Where small quantities of hot water are required out of normal hours in some areas, additional electrical non storage heaters may also be included.

In commercial premises and schools where separate space heating and D.H.W. boilers are installed the boiler plants can be economically used throughout the year.

Where domestic hot water is distributed from central storage calorifiers the the distribution losses can be very significant and in some schools the fuel from the D.H.W. boiler has been found to account for 40% of the total school fuel consumption. The lowest fuel consumption has been found to be associated with the "dead end" system where the tap must be run, to drain the cold water from the main before receiving hot water. Local By-Laws now prevent long runs of dead end water pipework being installed, but in many older primary schools they are a common feature. They are obviously very wasteful in water and do not give a satisfactory hot water service. With gravity circulating systems the standing losses are very high due to the large bore pipes used and circulation usually operates continuously even outside usage hours. Correctly designed pump return small bore systems are usually the best compromise and can be accurately time-switch controlled. To show the effect on the costs of operating the alternative systems the following data is based on field tests for typical commercial premises of 5,000m^2 (54,000 sq. ft.) of floor area, water being supplied to 6 toilet blocks and canteen from 7.0 a.m. to 6.0 p.m. five days per week. The costs per annum are:-

(a)   Central boiler plant (oil fired at 30p/Purchased   Therm)
    (i)   Dead end system   £810*
    (ii)   Gravity circulation   £1740
    (iii)   Pumped system with time-switch control   £1350
    (iv)   As for (iii) but with spray taps in basins   £1080
(b)   Central electrically heated calorifier   £2300
    (Pumped system with time switch and electricity
    at an average cost of 3.0p/unit)
(c)   Local electric heaters (electricity 3.0p/unit)   £1530
    ✻(not truly comparable due to lower standard of
    service, and additional water costs)

The temperature of water storage and type of taps etc., can also have a significant effect on water usage. Where standard wash basins and taps are used high temperatures necessitate the use of the plug and cold water addition so wastage is minimised.

With washing fountains the correct temperature to avoid scalding must be used so the important features are the volume of water flowing and tight shut off of the water flow when not required. With all domestic hot water applications the conservation of water and heat energy are complementary.

149

Spray taps can reduce hot water usage by over 50% and satisfy many people who prefer to wash under running water. Their cost is usually repaid within a year of installation, but there are some restrictions to their use, as hot and cold water supplies to them must be at similar pressures. This may mean that cold water should be supplied from a header tank rather than direct from town main.

ELECTRICAL SERVICES

Economies in useage not only assist in the general conservation of energy but also assist in effecting saving in the cost of purchased or generated electricity.

A much wider outlook is required however to effect overall cost savings; not just the directly visible one of electricity costs but also the indirect cost of providing and maintaining electrical services throughout the premises.

All too often we see industrial premises which have expanded over the years but insufficient thought has been given to the electrical distribution system due generally to misconceived ideas of capital costs. Alternatively new factories with ill planned electrical installations, have been built, not always the fault of the original electrical designer, but very often caused by lack of communication in respect of plant layouts and plant loads.

By having well planned electrical systems with in built allowances for future growth, the cost of electricity can be minimised by reduced heat losses in cables, which consume electricity, and also maintenance costs, which affect production.

Economies can also be affected by planned maintenance systems which assist in reducing breakdown times and consequential loss of production.

In almost every factory temporary wiring can be seen to at least one item of plant which has probably now become regarded as permanent but nevertheless remains as a high source of danger with possibilities of lethal consequencies. This type of temporary installation invariably is a result of poor liaison between production planning and maintenance

150

staff where insufficient time is given for permanent installations to be carried out for new plant or revised plant layouts.

In planning an industrial installation there are a number of items which require careful consideration, as the type of system and voltage of supply required will depend upon the various forms in which electrical energy is to be utilised in the factory and also on the total load requirments.

DISTRIBUTION

Having determined the total load requirments, arrangements can be made with the Electricity Board for the provision of an electricity supply.

The average medium sized factory will require a medium voltage supply of 415/240 volt 3 phase 4 wire 50 Hz which may well be provided from the Electricity Board's H.V. network via a transformer. The total load and tariff chosen will determine whether the transformer and associated protective H.V. switch gear will be owned by the Electricity Board or by the consumer. In any event a specially constructed area will be required, to house this equipment, and known as a sub-station.

Factories having large electrical loads may well require more than one transformer in which case a H.V. supply would be provided by the Electricity Board at an agreed position on the site and the Consumer would be responsible for providing the transformers and associated H.V. distribution on the factory site.

Transformers should be located as close as practicable to centres of heavy loads so as to minimise capital costs and losses associated with long runs of M.V. cables.

Transformers, associated H.V. switchgear and main M.V. switchgear should preferably be located adjacent to each other in separate chambers.

From the main M.V. switchboard, cable feeders would radiate to sub-main switchboards at local load centres to which would be connected final circuits for machines and distribution boards for lighting and small power.

Good practice and the safety requirements of the Factories Act and I.E.E. Regulations should be strictly observed when

151

Incoming H.V. supplies
from Electricity Board

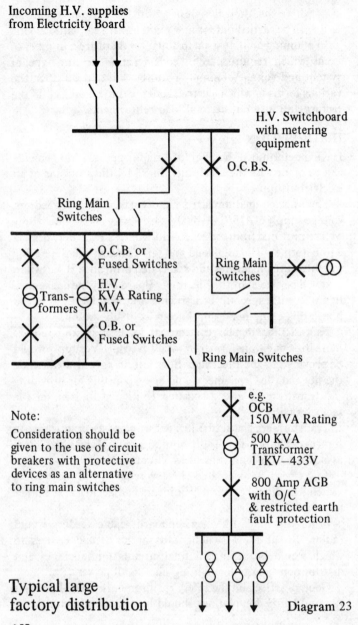

H.V. Switchboard
with metering
equipment

O.C.B.S.

Ring Main
Switches

O.C.B. or
Fused Switches

H.V.
KVA Rating
M.V.

Trans-
formers

O.B. or
Fused Switches

Ring Main
Switches

Ring Main Switches

e.g.
OCB
150 MVA Rating

500 KVA
Transformer
11KV−433V

800 Amp AGB
with O/C
& restricted earth
fault protection

Note:

Consideration should be
given to the use of circuit
breakers with protective
devices as an alternative
to ring main switches

Typical large
factory distribution

Diagram 23

152

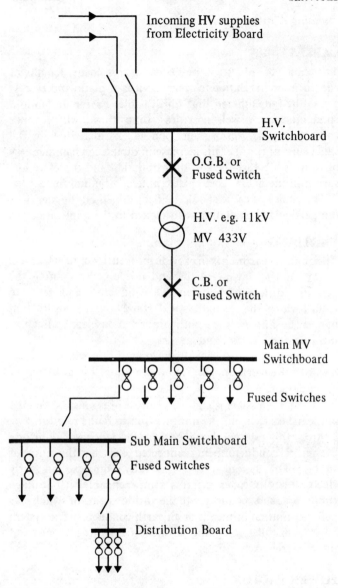

Incoming HV supplies
from Electricity Board

H.V.
Switchboard

O.G.B. or
Fused Switch

H.V. e.g. 11kV

MV 433V

C.B. or
Fused Switch

Main MV
Switchboard

Fused Switches

Sub Main Switchboard

Fused Switches

Distribution Board

Typical medium sized
factory distribution

Diagram 24

153

planning distribution systems.

CABLES

The choice of cables will be dependant on loads, lengths of run, location in relation to other services or plant, and cost. It should be remembered that whilst cables having aluminium conductors may well cost less than those with copper conductors, the overall dimension will be increased for the same current rating. This increase in dimension can and very often does increase the cost of installation and therefore any saving in the actual cost of the cable can be nullified.

The choice of cable should be determined on the merits of the particular installation with regard to the application.

EARTHING

The object of earthing is to provide for security of the electrical system and to safeguard life and property by limiting the potential difference between live conductors and earth to values which the insulation will stand and to ensure that dangerous potentials are not maintained on electrical equipment or non-electrical metalwork.

A return path for earth faults must be provided without hazards of electric shock, fire or explosion, by means of an earth circuit.

Distribution systems and metal enclosures must be bonded and earthed to avoid the dangers due to faults to earth.

An earthing system should be provided which would include the neutral conductor being connected to earth at the source of supply. The system would generally comprise an early electrode or electrodes or, if this is impracticable earth leakage circuit breakers, connected to the various items of switchgear and distribution by means of an earth conductor. The system must be installed to comply with the I.E.E. Regulations and the Factories Act.

POWER FACTOR

Power factor is the percentage of current in an AC circuit which can be used as energy, and is the ratio of true power, in

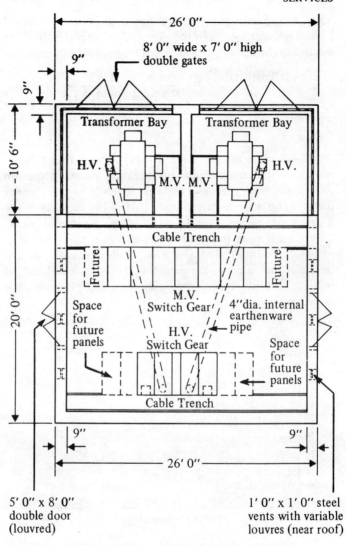

Typical twin transformer sub-station   Diagram 25

watts, to apparent power in VA. Apparent power is obtained by multiplying the volts by the current flowing in the circuit or by adding the active and reactive components of the current.

$\therefore$ Power Factor (P.F.) $= \dfrac{\text{true power (kW)}}{\text{apparent power (kVA)}}$

$$= \dfrac{\text{kW x 1,000}}{\sqrt{3} \text{ x V x A}}$$

All inductive equipment such as motors, transformers choke for fluorescent lights and welding plant operate on electro-magnetic principles and consequently have low power factors. The reactive current uses part of the capacity of the distribution network although it does no useful work. Therefore at peak times the Electricity Board's system may well become overloaded and it is for this reason that cost penalties are imposed through the tariffs, either direct or indirect, as a means of discouraging the use of plant with low power factors.

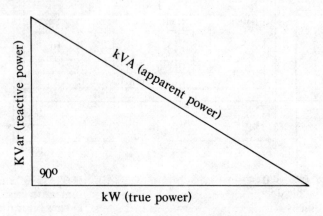

Relationship between Kw, kVA and kVAr.

Diagram 26

Where the power factor is excessively low, equipment can be installed to achieve connection to an economic level. This level

156

will be determined by the tariff on which the cost of purchasing electricity is based.

Improving power factor is not just applicable to purchased electricity but also where a factory has its own generating equipment, as by making such improvement increased useful power will be made available and thus improve the efficiencies of both generated and purchased supplies.

Example 1

A factory is supplied via a 1,000 kVA transformer at 425 volts between phases, the current taken is 1,040 amps and the measured power demand is 577 kW.

$$\text{kVA} = \sqrt{3} \times 435 \times 1,040$$

$$\text{Demand} = 764.7$$

$$\therefore \text{PF} = \frac{577}{764.7} = 0.75$$

By improving the power factor to 0.95 the kVA demand would be reduced to:-

$$\text{kVA} = \frac{\text{kW}}{\text{PF}}$$

$$= \frac{577}{0.95}$$

$$= 607.4$$

$\therefore$ There would be additional spare transformer capacity of (764.7—607.4) $= 157.3$ kVA thus allowing for more expansion at the factory without having to provide costly transformer and associated equipment.

In order to minimise the use of costly P.F. correction equipment when motor driven plant is utilised, the size of the electric motor should be carefully selected by closely matching the horse power of the motor to the mechanical load. Under

157

loaded motors and machines left idleing produce low power factors.

Power Factor Improvement

Generally there are two methods of improving power factor:-
(1) by the use of rotary equipment e.g. synchronous motors
(2) by the use of static equipment e.g. capacitors.

Either method has particular applications, the use of synchronous motors is the obvious method in some cases and in others capacitors will be more ideal. Each scheme must be considered on its merits, but it is not possible to make a direct basis for comparison.

Capacitors are used for improvement in a wide variety of industrial applications and this method is reliable, practical and economial.

The capacitors can be used on medium voltage systems, connected to individual motors, distribution centres or to the bus-bars of the main switch gear. High voltage capacitors are available for use where considered necessary. Whether bulk correction or individual correction is to be utilised, is again a case for the schemes to be treated on their merits as each scheme offers many advantages.

The overall power factor for the premises should never be connected such that a leading power factor, i.e. exceeding unity, is obtained. A leading power factor may well result in higher voltage on the factory distribution system thus accentuating voltage fluctuations and reducing lamp life.

Savings on electricity charges

Most Electricity Boards include in their industrial tariffs, some form of penalty for poor power factors. The tariff structures in general use are the kVA maximum demand tariff and the kW maximum demand tariff with average power factor penalty.

By installing correction equipment, savings in cost of purchased electricity can be achieved and the following examples give an indication of such savings associated with each type of tariff.

158

Example 2

A factory is charged for electricity on an annual kVA maximum demand tariff. The demand charges are £9.50 for each of the first 200 kVA of MD and £9.00 each of the next 300 kVA of MD. The factory's maximum demand is 450 kVA and the power factor is 0.80 and it is contemplated improving the power factor to 0.98.

Existing demand charges
200 kVA at £9.50	=	£1900.00
250 kVA at £9.00	=	£2250.00
annual demand charge	=	£4150.00

450 kVA at 0.80 p.f.	=	360 kW
kVA demand at o.98 p.f.	=	$\dfrac{360 \times 100}{98}$
	=	367 kVA

∴ Reduction in kVA demand = 450 — 367
= 83 kVA

∴ Annual saving to be set against cost of correction equipment would be

83 x £9.00 = £747.00

Example 3

A factory is charged for electricity on an annual kW maximum demand tariff and the demand charges are increased by 1% for each complete 1% by which the power factor falls below 90%.

The demand charges are £10.48 for each of the first 200kW of MD, £10.18 for each of the next 300kW of MD and £9.88 for each of the next 500kW of MD.

The annual maximum demand is 900kW and the average power factor is 0.75.

Existing demand charges
200kW at £10.48	=	£2096.00
300kW at £10.18	=	£3054.00
400kW at £ 9.88	=	£3952.00
		£9102.00

159

Choose Motor Size to Match the Load
(Diagram of motor current against load)

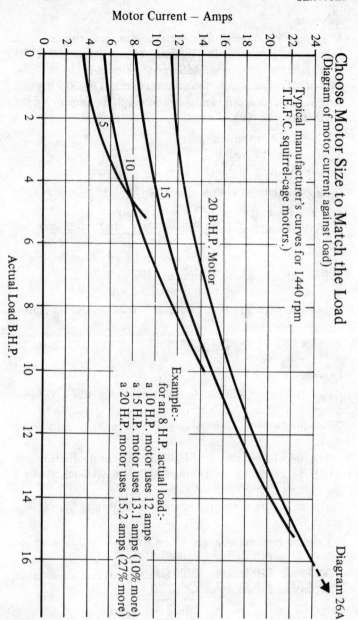

Motor Current — Amps

Actual Load B.H.P.

Typical manufacturer's curves for 1440 rpm
T.E.F.C. squirrel-cage motors.

5

10

15

20 B.H.P. Motor

Example:-
for an 8 H.P. actual load:-
a 10 H.P. motor uses 12 amps
a 15 H.P. motor uses 13.1 amps (10% more)
a 20 H.P. motor uses 15.2 amps (27% more)

Diagram 26A

160

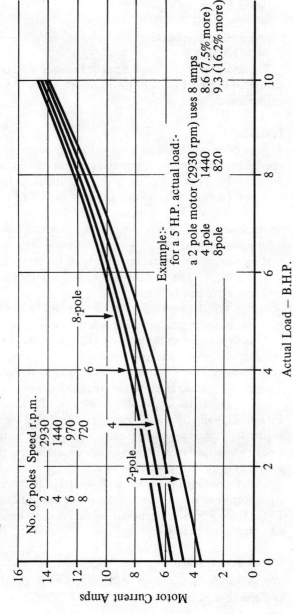

Use Highest Possible Motor Speed

(Diagram shows motor currents against load for different speeds of 10 B.H.P.
Three-phase induction motors.)

Diagram 26B

No. of poles	Speed r.p.m.
2	2930
4	1440
6	970
8	720

Example:-
for a 5 H.P. actual load:-
a 2 pole motor (2930 rpm) uses 8 amps
4 pole   1440   8.6 (7.5% more)
8pole   820   9.3 (16.2% more)

Actual Load – B.H.P.

Motor Current Amps

Power factor penalty
= 0.90 — 0.75
= 15% of £9102.00
= £1365.30

Therefore by improving the power factor to 0.90 the annual saving would be £1365.30 to be offset against the cost of correction equipment.

LIGHTING

Lighting design is a complex subject and required expertise in the field of illuminating engineering and detailed calculations on this subject have no place in this handbook. It is intended however, to show that by the prudent use of types of light sources available the most economical lighting installation can be provided.

In discussing lighting systems, it should be remembered that light, whether natural or artificial, is necessary in order to see. The degree and quality of light will determine just how well an object can be seen.

Such a broad statement is very easy to make but there are many factors to be considered in order to determine the degree and quality of light and any study of lighting systems embraces physics, physiology and psychology.

In designing any lighting scheme it is also necessary to comply with general statutory regulations and any other statutory regulation applicable to a particular industry. The Illuminating Engineering Society have produced recommendations for lighting building interiors, known as the I.E.S. Code, which includes levels of illumination for particular activities of the various industries.

The standards of illumination levels implied in the general legal requirements are lower than the levels indicated in the I.E.S. Code which are generally considered to be related to accepted good practice.

Visual Performance

Works production output will be affected by the visual performance of individuals.

162

The visual performance can be measured by means of the illumination which is necessary to perceive a particular size of detail or degree of contrast and also by the speed or the accuracy of carrying out a simple task.

Where illumination levels are low a significant increase in the visual performance will be produced if the levels are increased only moderately. However, whilst performance will continue to improve with more light, it will do so at an ever decreasing rate until a maximum value is reached when no further improvement will result from further increases in the illumination levels.

Freedom from glare is extremely important not only for visual comfort but also in carrying out the particular task. Common problems of glare within buildings are due to **direct glare**, caused by a poorly planned installation or the use of incorrect type of lighting fitting, and also **indirect glare** caused by reflections from shiny surfaces.

Light Source

In planning the lighting installation consideration must be given to the type of light source to be employed relative to the nature of the task and dimension of the area, together with the illumination necessary to carry out the task.

The light output from a light source is measured in lumens and illumination levels are measured in lumens per sq. metre.

The following table indicates typical light outputs from different light courses:

		lumens	lumens/watt
100 watt tungsten lamp	—	1160	12
150 watt tungsten lamp	—	1960	13
200 watt tungsten lamp	—	2720	13
500 watt tungsten lamp	—	7700	15
1500mm 65 watt fluorescent tube	—	4650	71
1800mm 85 watt fluorescent tube	—	6100	72
2400mm 125 watt fluorescent tube	—	8600	69
125 watt fluorescent mercury lamp	—	5500	44
250 watt fluorescent mercury lamp	—	12400	50
400 watt fluorescent mercury lamp	—	21800	55

250 watt high pressure sodium lamp — 21000   84
360 watt high pressure sodium lamp — 34500   96
400 watt high pressure sodium lamp — 40000   100

It will readily be seen from the above table that the maximum demands of a consumer are greatly affected by the type of light source employed.

Lighting Fittings

The choice of lighting fittings will be dependant obviously on the light source to be employed and also on the nature of the task and the building.

In choosing lighting fittings, consideration should be given to the location of the fittings in relation to the task, the mounting height, the nature of the task and the maintenance required.

Maintenance of lighting fittings and lamps is all too often ignored. Dirty lamps and fittings reduce the illumination level and regular cleaning will assist in maintaining the designed levels. Typical measurements show a 10-15% loss in lighting levels if regular cleaning is neglected.

Light output from lamps diminishes continuously with age and there comes a time when the extra light from new lamps justifies the capital cost and labour cost of replacement. A planned system of bulk lamp replacement will very often show a saving in the cost of maintenance labour.

General

Illumination levels are not only effected by the light source and fittings but also by the reflective surfaces of walls, floors and ceilings.

Cleaning and redecoration of walls and ceilings has an immediate effect on the illumination levels in addition to enhancing the appearance of rooms or factory areas.

If new lighting installations or modifications to existing are contemplated it is advisable to seek expert guidance so that the most economical system can be employed. By having the correct system, economies will be seen both in capital and running costs.

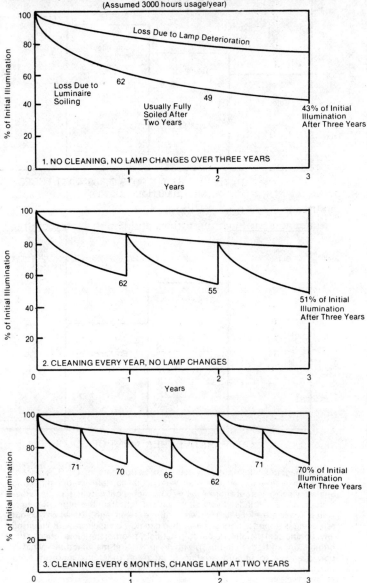

DIAGRAM 26C

EFFECT OF CLEANING REFLECTOR, ANY TRANSPARENCY, AND LAMP; ALSO RENEWING LAMP AFTER FIXED PERIOD
(Assumed 3000 hours usage/year)

Loss Due to Lamp Deterioration

Loss Due to Luminaire Soiling

62

49

Usually Fully Soiled After Two Years

43% of Initial Illumination After Three Years

1. NO CLEANING, NO LAMP CHANGES OVER THREE YEARS

% of Initial Illumination

Years

62

55

51% of Initial Illumination After Three Years

2. CLEANING EVERY YEAR, NO LAMP CHANGES

% of Initial Illumination

Years

71

70

65

62

71

70% of Initial Illumination After Three Years

3. CLEANING EVERY 6 MONTHS, CHANGE LAMP AT TWO YEARS

% of Initial Illumination

Years

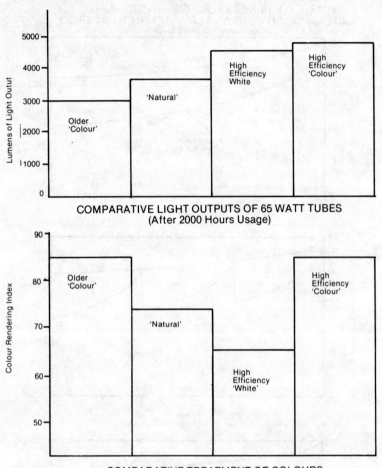

COMPARATIVE LIGHT OUTPUTS OF 65 WATT TUBES
(After 2000 Hours Usage)

COMPARATIVE TREATMENT OF COLOURS

These diagrams illustrate how recent developments in 6ft. fluorescent tubes have produced a tube with very good 'daylight' colour rendering and yet which gives out approximately 60% more light output than the earlier 'colour' tubes and indeed very slightly more light than the 'high efficiency white' tubes developed only two to three years ago. So where colour rendering is important, these new tubes can now be used without incurring any running cost penalty. Do not automatically re-order the same tubes as the original ones fitted. Find out by enquiry to several manufacturers what is now available to fit into existing fittings and order the more efficient tubes — even where capital cost is a little higher.

166

For example at one firm, NIFES were able to demonstrate that there was a case, both on savings of electricity and reduction of wastage by improving levels of illumination, to replace the existing old tungsten-filament bulb system by modern flourescent lamps and tubes. A programme of replacement was put in hand using small local contractors to carry out the work. However, and at the same time, the first were having an extension built by a national contractor and to everyone's surprise this included tungsten-filament lighting as part of the "turnkey project".

For further reading on the subject of lighting, the reader is recommended to the following books which can be obtained from the Electricity Council.

Electrical Services in Buildings—Electricity Council
Management Guide to Modern Industrial Lighting—
By Stanley Lyons, M.Illum.E.S.

7
Standby Consideration: Generation of Permanent Supply

INTRODUCTION—WHY GENERATE?

The public Electricity Generating Boards have an excellent record in fulfilling their statutory function of providing a dependable supply of electricity at the lowest possible cost to the consumer. A proportion of the fuel they burn is also very low grade, which could not be effectively used by industry with existing plant.

On the other hand, public electricity generation is based on the condensing steam turbine cycle. The overall CEBG power station thermal efficiency is around 32% and scope for improving this is marginal. Taking account of transmission losses between the power station and consumer, the power delivered to the factory represents approx. 28% of the heat content of the primary fuel burned at the power station.

This situation is unlikely to change appreciably in the near future. The long term policy of the Generating Boards is to rely on nuclear power stations to reduce substantially their demand for fossil fuels. This is, however, dependent on the development of a foolproof nuclear waste disposal system—we are still far from achieving this. Even if the Board's attitude and statutory limitations are altered and consideration is given to the installation of combined heat and power stations and large hydro and tidal schemes to husband the nation's primary energy resources, the lead time associated with such projects is such that no significant development will take place in the next two or three decades.

From the industrialist's point of view, a supply of electricity is indispensible. It represents over half industry's total energy consumption, neglecting specialised use of coal in coke ovens. It is also the most expensive form of energy used being at least 2½ to 4 times the cost of the other important forms of energy—coal, oil and natural gas.

Also, disruptions in the public electricity supply over recent years, although small in number and duration, have driven home to the industrialist the disproportionate resultant cost in lost production.

Because of the disadvantages of scale, industry in general cannot compete economically with the public Generating

171

Boards in the straight production of electricity.

If a particular industry does have both a power and heat requirement and these are in reasonable balance, an economic case can often be made for the combined generation of power and heat at an overall thermal efficiency of between 60% and 80%. In addition to the financial advantage to the industry concerned, this can make much better use of the nation's primary energy resources. This type of system has been traditional in such industries as paper and chemicals, and can be applied on a much wider scale. See Diagram (27).

A case for private power generation also exists where substantial quantities of waste heat are available, e.g. either low grade vapours from evaporators or high grade waste heat from furnaces.

In recent years there has been a trend in industry to install standby generating plant to safeguard essential production in event of failure of the public supply. The economics of this are essentially an insurance calculation, but the return on capital expended can sometimes be improved by running the generating plant over the year to generate a proportion of the works power requirements or to limit expensive peak demand costs on the purchased power supply.

There is no general case for industry to generate its own electricity requirements and each case must be carefully evaluated on its own merits. Where the conditions are favourable, the economic advantage to industry and to the nation can be substantial. Financially viable schemes for private generation may apply for about a quarter of the electricity at present purchased by industry, saving the national approximately 10 million tons coals equivalent/annum in primary energy usage. In the remainder of industry, detailed analysis is likely to reveal no economic case for private generation. Scale is not the sole criterion. Economic cases have been established in special circumstances for generating as little as 50-100 kW.

The reasons for industry considering private generation are therefore:-

a) Economic

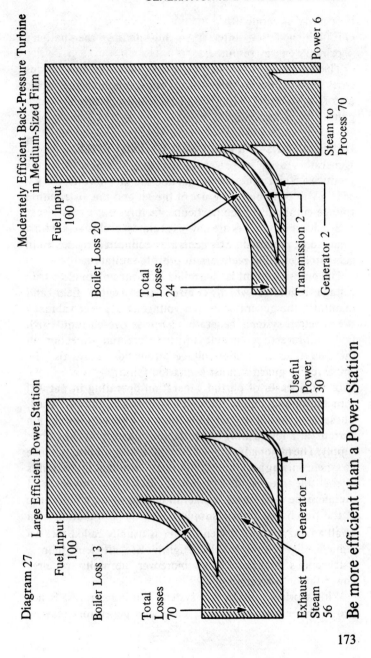

Diagram 27

Large Efficient Power Station

Moderately Efficient Back-Pressure Turbine in Medium-Sized Firm

Fuel Input 100

Boiler Loss 13

Total Losses 70

Generator 1

Exhaust Steam 56

Useful Power 30

Fuel Input 100

Boiler Loss 20

Total Losses 24

Generator 2

Transmission 2

Steam to Process 70

Power 6

Be more efficient than a Power Station

b) Security of supply
c) Enlightened self interest in husbanding the nation's primary energy resources.

The following three sections briefly describe the various types of machines in general use, and examine some of the factors pertaining to the installation and operation of such plant.

GENERATORS—TYPES USED BY INDUSTRY

Generators used in private generating schemes normally produce a 50 Hz A.C. supply and operate at 440 volts, 3.3 kV or 11 kV depending on the size of the set and the distribution voltage employed in the particular factory.

Brushless generators are now used almost exclusively, with a small shaft mounted A.C. generator connected to the main generator rotor via rectifiers to provide excitation.

The output current is controlled by means of an electronic voltage regulator which controls the exciter field and maintains the generator output voltage at a preset value.

Protection systems generally comprise overcurrent, earth fault and reverse power relays with differential protection on the larger sizes. Under-voltage protection may also be employed to guard against excitation failure.

In a few cases of partial generation operating in parallel with the public supply, induction generators may be used. These are similar to an induction motor with excitation current, at a low leading power factor, taken from the public supply. The major advantage of this system is simplicity, since no synchronising gear is required. A disadvantage is that if this is the sole means of generation, excitation current and therefore the whole private generation is lost in event of failure of the purchased public supply hence the installation of a small auxiliary petrol driven exciter is usually called for. An induction generator is also slightly less efficient than a synchronous alternator and moreover normally requires power factor correction gear.

With a private generation system, the industrialist is not tied to the public supply frequency, and generators may be installed operating at say 400 hz for lighting, or at high frequency for special process requirements, such as induction heating or radio frequency drying.

D.C. power is very occasionally generated privately, but generally this is dictated by the need to supply existing D.C. motor power drives or small loads, such as emergency lighting or battery charging without the complication of rectifiers. It is unlikely to be an economic proposition for new installations where any D.C. requirements can usually be met more conveniently by means of rectifying equipment operating from the normal A.C. system.

TYPES OF PRIME MOVER

The most common generator prime movers for industrial applications are steam turbines, diesel engines and gas turbines although petrol engines may be used on a very small scale. Although regarded by many as obsolete, high speed steam engines are still used to a limited extent for power generation and for local direct drives.

Steam Turbines

There are three basic types of steam turbine to generate power as a by-product of process of exhaust steam:-
a)'Condensing
b) Passout Condensing
c) Back pressure

These are manufactured for outputs from 50 kW to 30 MW. Most industrial installations are in the range 1-5 MW. The greater the steam pressure drop through the turbine the greater will be the power output. A reduction in the exhaust pressure effects a greater increase in power generation than an increase in inlet steam pressure. Specific steam consumptions depend on the absolute pressure ratio over the turbine and the size of set considered.

a) Condensing turbines tend to be high in cost, thermally inefficient—around 10-25% efficiency, and bulky relative to power output. They can generally only be used economically for private generation where steam can be generated from

175

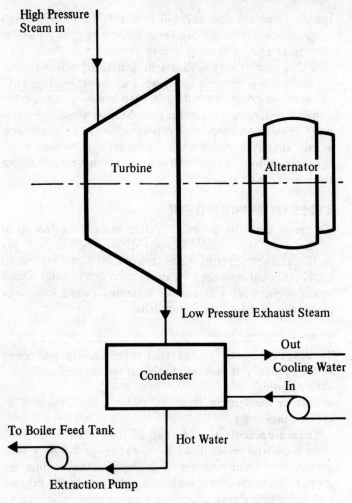

Condensing Turbine

Diagram 28

waste heat from some other process, or to utilise low presure vapours from evaporators or similar plant. As a result, they are never used for total generation schemes, but can, in favourable circumstances, be used to provide local mechanical drives to replace electric motors, or provide a proportion of works electrical power requirements. Inlet steam pressures could be anything from 650 psi down to atmospheric pressure or slight vacuum. Exhaust vacuum should be as high as is economically practicable. See Diagram 28.

b) Passout condensing turbines are the most common type for total generation schemes since, within limits, electrical output can be adjusted by altering the proportions of steam passing to the condenser and steam demand in excess of turbine passout capacity can be provided through a reducing valve and desuperheater bypassing the turbine. Inlet steam conditions can be in the range 650 lbf/in^2, 750°F down to 250 lbf/in^2 saturated. Exhaust vacuum is generally 28ins. Hg or greater. Cycle thermal efficiencies in the range 50-70% can be obtained but may be much lower if more than the minimum steam is passed to the condenser. See Diagram 29.

c) Back pressure turbines are inexpensive, thermally efficient and compact and generally the most economical proposition for partial generation schemes. Power generation is dependent on the steam flow to meet process requirements and the pressure drop over the turbine. The running cost for electrical generation is therefore the marginally additional cost of generating steam at higher pressure and temperature than would be required for process usage only, plus the fuel equivalent of the heat drop across the turbine. See Diagram 30.

In the simplest case a back pressure turbine can be used to replace an existing steam pressure reducing valve either generating electricity or replacing electric motors providing mechanical drives.

Exhaust pressure from the turbine is determined by process steam pressure requirements. Inlet steam conditions depend on power generation required. In industrial

177

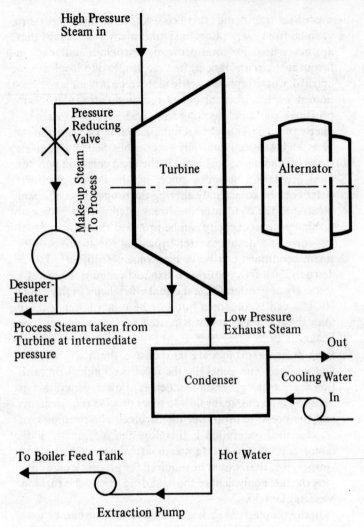

High Pressure
Steam in

Pressure
Reducing
Valve

Make-up Steam
To Process

Turbine

Alternator

Desuper-
Heater

Process Steam taken from
Turbine at intermediate
pressure

Low Pressure
Exhaust Steam

Out

Condenser

Cooling Water

In

To Boiler Feed Tank

Hot Water

Extraction Pump

Passout Condensing Turbine

Diagram 29

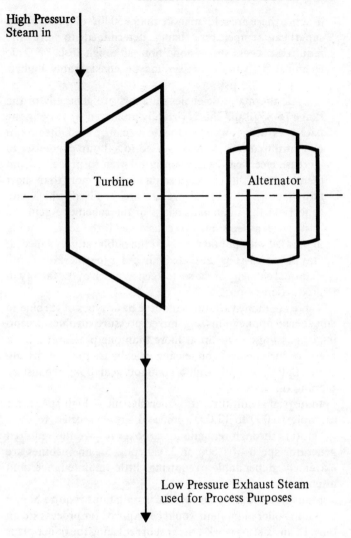

High Pressure Steam in

Turbine

Alternator

Low Pressure Exhaust Steam used for Process Purposes

Back Pressure Turbine

Diagram 30

practice these are seldom more than 650 lbf/in^2, 750° F, the superheat temperature being determined to give the required process steam condition, although in large scale chemical industry processes maybe considerably higher, e.g. up to 1900 psi.

Cycle thermal efficiencies obtained are generally in the range 75-85% and this is virtually unaltered by varying the back pressure provided all the steam can be used. Steam can be blown to atmosphere or passed to a dump condenser to increase electrical generation in relation to process steam demand but if this is practiced on anything more than short term emergency excessive costs are involved which can cancel out the financial savings of the scheme. Again the 'insurance' aspects must be considered if the set is running in parallel with public supply. If the public supply failed, it may pay to waste exhaust steam in order to generate the full electrical capacity of the set to keep as much of the factory in production as possible.

There are many variations of these basic types of turbine to suit special applications, e.g. mixed pressure turbines accepting or passing out steam at more than one pressure; a back pressure turbine and condensing or exhaust turbine driving the one alternator through a common gear box; exhaust or topping turbines.

Industrial steam turbines generally run at high speed, for example 10,000 to 12,000 rpm, and are connected to their generators through reduction gearboxes to give the required generator speed of 1,500 or 3,000 rpm. Steam turbines are extremely dependable, requiring little maintenance and attention.

Many industrial steam turbine installations require higher pressure boiler plant than would be required for process steam only. In an existing works, the cost of replacing the boiler plant may make a power generation scheme uneconomic. When considering a new factory installation, or if a power generation installation is timed to coincide with replacement of an existing process boiler plant, only the difference in cost between new LP and HP boiler plants need be related to the saving in purchase power costs. In these circumstances, the scheme is much more likely to be an economic proposition.

Diesel Engines

Into this category come engines running on light fuel oil, heavy fuel oil and natural gas. Dual fuel arrangements can also be obtained—generally using natural gas and light oil.

The common size range of diesel generators is 30 kW to 1,000 kW but both smaller and larger sets are readily available. These nominal ratings are reduced in relation to the height of the installation about sea level, and ambient temperature above 30°C.

There is also a range of engine operating speeds available from high speed, around 1,200 to 1,800 rpm, to low speeds below 400 rpm. The less expensive high speed engines, mainly in the smaller sizes and running on light oil, tend to be the most economic proposition for standby duties. The more expensive lower speed engines often running on heavy oil, tend to be selected for larger units and continuous operation because of their higher efficiency, greater life and lower maintenance requirements. Larger engines are usually pressure charged and fitted with air coolers, since this can increase power output by up to 50% and improve efficiency by up to 4%.

Most diesel engines operate on the four stroke cycle, but in the larger slow speed range there are some very dependable two stroke units.

Specific fuel consumption varies with engine size, design and imposed loading. Large turbo charged medium speed sets in the 1,000 Kw output range can consume around 0.22 Kg-0.25 Kg light fuel oil per kW output per hour at full load, rising to 0.24 Kg-0.27 Kg per kWh at half load. Performance below half load deteriorates rapidly and it is normal to run engines at as high load factor as possible. See Diagram (31)

Thermal efficiency as a straight electrical generator can vary from as low as 25% at full load in the 30 kW size to as high as 39% at full load for sizes above 1.500 kW. Considering the relatively high cost of fuel and maintenance, this cannot

181

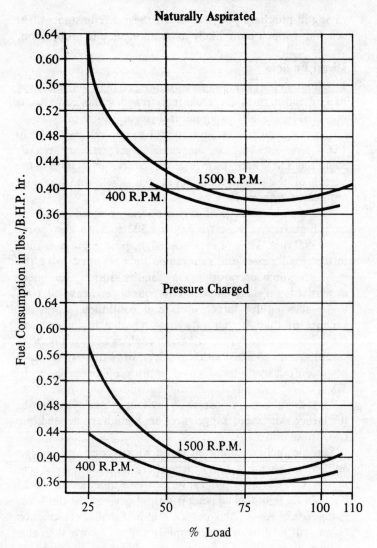

Naturally Aspirated

Pressure Charged

Fuel consumption of Diesel Engines   Diagram 31

compete economically with the purchased power supply unless other factors, such as safeguarding production in event of public supply failure, are also taken into account.

Of the total heat in the fuel supplied, approximately 30% is rejected in the exhaust gases and between a half and two thirds of this can be readily recovered by heat exchangers such as waste heat boilers to provide hot water or steam to the factory services. Also, some 18% of the heat output is dissipated by jacket and lubricating oil cooling water. Nearly all this is recoverable by suitable heat exchangers, but it is not always economic to do so. This potential heat recovery can raise the overall thermal efficiency of the plant to between 40% and 70% provided full use can be made of the recovered heat. Heat recovery, however, falls rapidly with reducing engine load and in the case of steam generation with increasing steam pressure requirement. See Diagram (32) to Diagram (33)

On any medium to large size installations, say of over 500 kW installed capacity for continuous duty it is always worth considering the additional capital cost of full heat recovery against the overall running cost savings. In some cases, particularly on heavy oil, a combined power and heat scheme can compare favourably with purchasing electricity and generating heat requirements independently.

With all installations, adequate provision of combustion and cooling air is essential. A cooling water system with cooling towers (or heat exchangers if river water can be circulated) is generally required. In most continuously operating installations and many standby plants, consideration has to be given to acoustic treatment of foundations and buildings or acoustic enclosures.

Petrol Engines

Petrol engine driven generating plant is available, but is generally restricted to outputs of up to 100 kW. In industrial practice their use is usually limited to small compact portable units of up to 10 kW output for emergency lighting, battery charging duties, or to enable larger plant to be started up after a power failure. The high engine and fuel costs generally

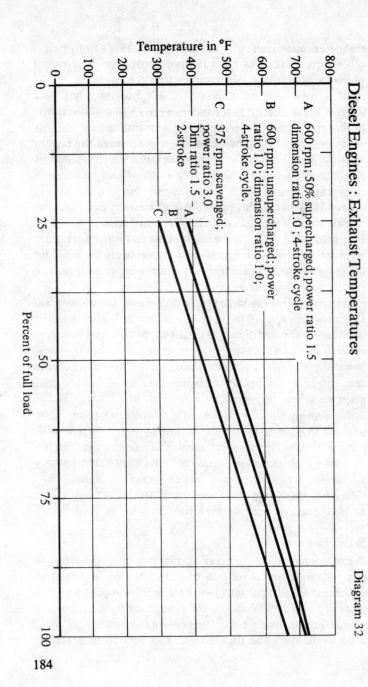

Diesel Engines : Exhaust Temperatures

Temperature in °F

A 600 rpm; 50% supercharged; power ratio 1.5
   dimension ratio 1.0 ; 4-stroke cycle

B 600 rpm; unsupercharged; power
   ratio 1.0; dimension ratio 1.0;
   4-stroke cycle.

C 375 rpm scavenged;
   power ratio 3.0
   Dim ratio 1.5
   2-stroke

Percent of full load

Diagram 32

184

Showing waste heat recovery from diesel engines

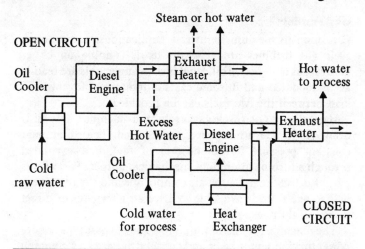

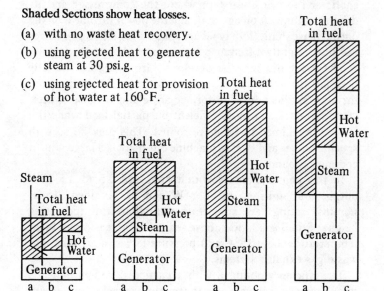

Shaded Sections show heat losses.

(a) with no waste heat recovery.

(b) using rejected heat to generate steam at 30 psi.g.

(c) using rejected heat for provision of hot water at 160°F.

Heat Loss v Generator Load

Diagram 33

185

preclude their consideration in the larger size ranges.

Gas Turbines

For small to medium industrial applications, simple open cycle gas turbines are generally used, running on either natural gas or light fuel oil. Dual fuel arrangements are readily accommodated and in some cases automatic changeover on load between the two fuels can be provided.

Basically the open cycle gas turbine is a machine where air is drawn from the atmosphere, compressed, heated at near constant pressure by combustion of fuel, and expanded through a turbine to produce mechanical output. Part of this output is used to drive the air compressor and the remaining half to one third is available to drive an alternator or direct mechanical drive.

The machine may be of the simple single shaft type; where power turbine, compressor and output drive are on a common shaft; or the twin shaft type where the compressor and its driving turbine are on one shaft and the power turbine is on the output drive shaft. Both types are suitable for alternator drives but have slightly different power control characteristics.

The machines may be developed from engines initially designed for aircraft application or units specifically designed for land application. The former are claimed to be lighter, more compact and more efficient on partial load while the latter are claimed to be more robust. This may be so with specific makes and sizes of machine but there is a large area of common ground.

In this country, common unit sizes are 0.5 to 3.5 MW output, but units up to 60 MW can be obtained. These nominal ratings are subject to reduction with increasing height of the installation above sea level, increasing ambient temperature about 15°C and pressure losses in air inlet and waste gas exhaust systems.

The efficiency obtained with a particular design depends' upon the cycle pressure ratio, the maximum cycle temperature and the aero-dynamic efficiency of compressor and turbines. In the small industrial size range electrical generation

186

efficiency can be as low as 14 to 20% at full load, although larger more sophisticated units can achieve electrical efficiencies in excess of 30%. Specific fuel consumptions of a typical 1.25MW set can vary from 0.51 Kg gas oil per KW hour at full load to 0.63 Kg per KW hour at half load at an ambient temperature of 15°C, increasing however with higher ambient temperatures particularly on partial load. See Diagrams 34 and 35.

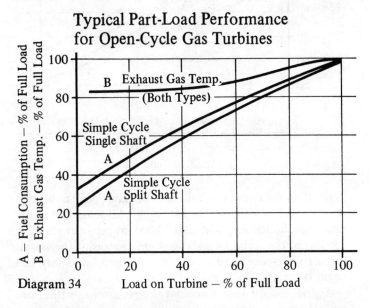

Typical Part-Load Performance for Open-Cycle Gas Turbines

Diagram 34   Load on Turbine — % of Full Load

Virtually all the total heat input in fuel not converted to shaft power is available for heat recovery.

Radiation and oil cooler losses only amount to 1 to 2%. The waste gas is clean, nearly all air, and at high temperature—over 500°C on full load. Exhaust gas mass flow reduces only slightly on falling load, but gas temperature can fall to about 420°C at half load reducing heat recovery potential.

If the waste gases were passed through a waste heat boiler to generate steam, the 1.25MW set could generate some 10,000 lb/hour at 250 lbf/in^2 raising the combined electrical and steam generation efficiency to around 50%.

187

Effect of Temperature and altitude on permissable continuous output of typical open-cycle gas turbine.

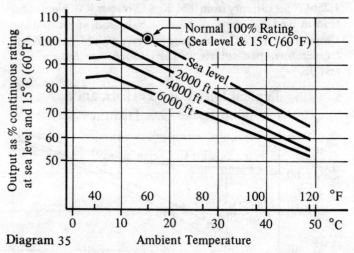

Diagram 35   Ambient Temperature

If additional steam is required, the exhaust gas can be used as preheated combustion air for firing additional fuel in the waste heat boiler. An additional 30,000 lbs per hour steam at 250 lbf/in^2 could comfortably be raised, increasing the overall thermal efficiency to over 70%, also the supplementary fuel could be heavy oil.

As an electrical generator alone the gas turbine is therefore less effective than the diesel engine. Its waste heat recovery potential is, however, greater and it is more flexible.

The diesel engine would therefore tend to be used where the electrical demand is high in relation to steam or heat demand, say less than 3kg(7lb.) steam per kW of electricity load, and the gas turbine cannot develop to full thermal efficiency potential unless more than 4 kg (9 lb) of steam is required per kW of electricity.

These figures compare favourably with a straight back pressure steam turbine, where with typical medium industries' ranges of inlet and process steam pressures, the steam load must exceed 10 kg (22 lb) per kW of electricity load. See Diagram (36).

As an alternative to steam generation the waste gas may be used direct or through a heat exchanger for many industrial drying applications.

Gas turbines have many advantages over diesel engines for combined power and heat installations. They are low in weight and operate with little vibration requiring inexpensive foundations. Maintenance costs are low and dependability is high. Also no cooling water system is usually required, since a simple fan blown air cooler is generally sufficient for oil cooling.

The initial cost tends to be higher than diesel engines, although this is generally offset by lower running costs for continuously operating plant. A problem with gas turbines can be noise emission. Even small units in the 1-3 MW range can emit sound pressures around 100 dBA at one metre although this is mainly in the high frequency range—4,000 cycles and can readily be attenuated. The most common approach is to fit acoustic enclosures round the machine itself, or to house each machine in a solid construction acoustic cell. In either case, silencers are required on air inlets and cooling air exhaust. The heat recovery equipment and chimney often provide sufficient attenuation for exhaust gas sound emission.

PARELLEL, INDEPENDENT OR SECTIONED OPERATION

When a firm considers installing private generating plant to operate continuously, as opposed to emergency standby equipment, there is the option to operate in parallel with the public electricity supply. This can apply whether it is intended to generate the total works electricity requirements or only a portion of them.

Parallel Operation (Diagram 37)

Private generating plant operated in parallel with the public supply is in effect an extension of the national network. As

189

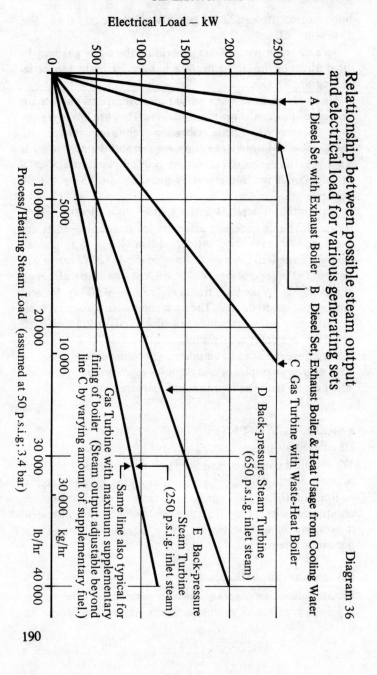

Electrical Load – kW

Relationship between possible steam output and electrical load for various generating sets

Diagram 36

A Diesel Set with Exhaust Boiler   B Diesel Set, Exhaust Boiler & Heat Usage from Cooling Water

C Gas Turbine with Waste-Heat Boiler

D Back-pressure Steam Turbine (650 p.s.i.g. inlet steam)

E Back-pressure Steam Turbine (250 p.s.i.g. inlet steam)

Gas Turbine with maximum supplementary firing of boiler (Steam output adjustable beyond line C by varying amount of supplementary fuel.)

Same line also typical for

Process/Heating Steam Load (assumed at 50 p.s.i.g.; 3.4 bar)

lb/hr   kg/hr

190

Two Generators arranged for running parallel with the National Grid

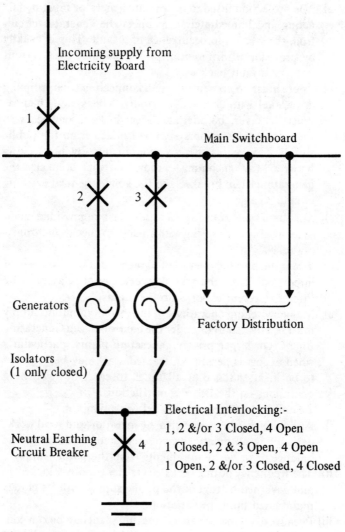

Incoming supply from
Electricity Board

1

Main Switchboard

2   3

Generators

Factory Distribution

Isolators
(1 only closed)

Neutral Earthing
Circuit Breaker   4

Electrical Interlocking:-
1, 2 &/or 3 Closed, 4 Open
1 Closed, 2 & 3 Open, 4 Open
1 Open, 2 &/or 3 Closed, 4 Closed

Diagram 37

191

such, the standards of design, installation and operation must satisfy the Electricity Boards that they will, in no circumstances, endanger the Board's system. Some of the essential requirments are:-

a) The protection fitted to the private generator must be fast acting and immediately disconnect the generator circuit from the system on occurrence of a fault. This is usually by provision of differential protection with over-current and earth fault back up.

b) There must be no neutral earth connection when running in parallel with the public supply. The system neutral must, however, be adequately earthed on any occasion the private generation system is isolated from the public supply. This requirement is generally met by interlocking between the incoming supply circuit breaker, the generator circuit breaker and the system neutral earthing circuit breaker.

c) Adequate synchronising facilities must be provided, such as manual synchronising with a back up check synchronising relay.

d) When the parallel operation agreement is on the basis of import only from the grid, a reverse power relay must be fitted to prevent export into the mains.

e) Major switching operations of the incoming public supply must be effected by or under the control of, the Generating Board. On larger private generating plants, particularly when export is permitted, trained staff may be required to be in attendance at all times to carry out switching operations on the Board's instructions.

The chief advantages of parallel operation are:-

i) Any deficiency between private generation and total works demand is automatically compensated by drawing the balance of power requirements from the public supply.

ii) The frequency of the works electricity supply is locked to and governed by that of the public supply with its closely maintained limits of variation.

iii) Because of (a) the private generating plant can be run continuously at full load or at its most economic overall load factor, to achieve minimum works costs taking into account purchased and works generated power costs and waste heat recovery.

iv) In event of failure of the works generating plant, the public supply will provide an automatic standby subject to any faulty plant being disconnected from the system. This avoids the capital and maintenance costs of private standby generating plant.

v) In event of failure of the public supply the private generating plant can still be operated within the limits of its capacity, provided means of starting essential auxiliary equipment have been installed.

vi) Where the parallel running agreement includes for exporting power, private generation surplus to works requirements can be automatically exported to the public supply network. Should this export be substantial, and over long periods, a small unit price for exported kW payable to the industrialist can be negotiated with the public supply authority.

Against these advantages certain costs will have to be weighed. The industrialist will be required to pay the Electricity Board an annual charge related to the standby capacity element of the Board's connected supply and related equipment. This will vary from Board to Board, but is generally based on the installed or normally operated rating of the private generating plant, and may be at a rate of £4.5 to £7/annum per installed kVA.

Also if this standby capacity has to be used, due to private plant failure or shutdown for maintenance for example, the purchased power over this period will be charged at full maximum demand tariff or other special high rates written into the agreement for parallel operation.

If power is exported only minimal rates are paid to the industrialist generally based on the Generating Board's fuel costs. It is not therefore economic to export power to the public supply except on a very large scheme with preferential agreed tariffs or other very special circumstances.

Independent Operation

In some cases when a firm can generate all it own power requirements from process steam or similar "total energy" basis, there may be no need to operate in parallel with the public supply or even have a public electricity supply at all.

In such circumstances, independent operation can be the most economical method of power generation and is common practice in the paper making and chemical industries.

Major considerations are the number of generating units to install to meet the total power requirements and the extent to which standby plant should be installed to cater for essential maintenance or unprogrammed failure of generating units and essential auxiliaries. This will be considered in more detail under reliability.

If production is to be shut down every weekend, it may not be economical or practical to generate the small amount of electricity required for essential lighting and maintenance operations over this period. A small public electricity supply could then be retained serving segregated power and lighting circuits, enabling the generating plant to be shut down. If the works is served by a small number of large boiler units, it may also be economic to have a small standby boiler to provide essential heating services over the weekend shutdown period.

Consideration has also to be given to the power requirement for starting up the main boilers and generating plant after the shutdown. This could be met by a small diesel generator or in the case of a gas turbine installation, either A.C. motor start from the public supply, compressed air start with a diesel driven compressor, or D.C. start from a trickle charge storage battery system.

Sectioned Operation

When only a proportion of the works electricity requirements is to be supplied by private generating plant, sectioned operation is the alternative to parallel operation.

In its simplest form the plant to be served from the private generating equipment can be on completely independent circuits from the remainder of the works served by the public supply. Purchased maximum demand and units will be reduced to the extent of private generation. This reduction will, however, be limited to the actual usage of the connected equipment which may be less than the potential generating capacity. The dependability of the independent section will also be less than the remainder of the works served by the public supply.

If it is required to increase the dependability of the private generation section by using the public supply as a standby, the public and private supplies to the switchgear serving the plant must be connected through interlocked circuit breakers to make it impossible to connect the public supply until the private supply is isolated and vice versa. With such an arrangement, the Electricity Board will generally apply standby charges as in the case of parallel operation.

In general, sectional operation can be lower in first cost than parallel operation, and is often used when private generation is only a small part of total electrical requirements and slightly lower dependability than the public supply can be accepted. As the extent of private generation increases and if it is desired to retain the public supply as standby, the economic balance moves in favour of parallel operation.

IMPLICATIONS OF STANDBY REQUIREMENTS

The main reasons for installing generating plant for standby duty only are to provide a supply of electricity to enable production processes to be continued or to prevent damage to or loss of product or manufacturing plant in the event of an interruption of the public electricity supply.

The economics of standby generation are essentially an insurance type calculation. The cost of probable loss of production or product and all associated costs due to an infrequent and probably short term failure of public electricity supply, must be weighed against the capital, maintenance and running costs of the necessary equipment to minimise such loss.

Factors such as non productive wages and external considerations, such as the ability of suppliers to continue to

195

supply raw materials or customers to accept supplies over a prolonged failure of public electricity supply have also to be taken into account.

Since there is no guaranteed return from the capital outlay involved and the actual running time of installed plant is unlikely to be more than a few weeks per year, low cost self contained generating units which can fit into existing factory space are favoured. This generally means high speed diesel generators or in special cases gas turbines.

Also, it is seldom economic to consider generating the total works electrical requirements on a standby basis. The capacity of installed plant is generally limited to that necessary to power certain essential process loads and emergency lighting, which could be at most, three quarters of normal requirements and generally very much less.

The standby generators may be connected to the main factory switchboard, enabling generated power to be distributed throughout the works. On large installations where there are a number of switchboards, generators feeding into selected switchboards are sometimes employed. Whichever system is adopted however, it is necessary to ensure that the switch on the Electricity Board's supply is opened before starting the standby plant and connecting it to the switchboard and conversely isolating the standby connections prior to reconnecting the public supply.

The necessary interlocking arrangements must be agreed with the local Electricity Board. (Diagram 38)

The time delay which can be tolerated between a failure of the public supply and the connection of the standby supply, largely determines the type and cost of the changeover system selected. There are three basic options:-

a) Manual—this is the most common and lowest cost system. On failure of the public supply the incoming circuit breakers are manually opened, releasing the interlocks to enable the standby plant circuit breakers to be closed. Distribution breakers are opened, the standby plant is then manually started, the selected generator and distribution circuit breakers closed and motors etc. throughout

Single Generator arranged for standby operation only.

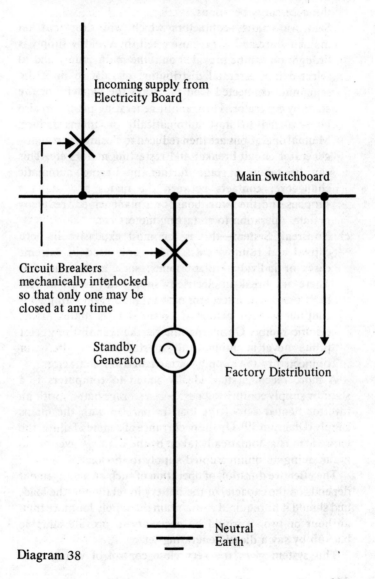

Incoming supply from
Electricity Board

Main Switchboard

Circuit Breakers
mechanically interlocked
so that only one may be
closed at any time

Standby
Generator

Factory Distribution

Neutral
Earth

Diagram 38

the works restarted. Depending on the extent of the installation this operation can take from say ten minutes to half an hour, provided staff is always available to carry out the necessary operations.

Semi-Automatic—contactors which will drop out on mains failure and not reclose when the standby supply is brought on can be provided on i) the main suuply and ii) selected non essential distribution circuits to limit the remaining connected load to within the capacity of the standby generator. The standby generating plant can also be arranged to start automatically on mains failure. Manual operations are then reduced to closing the standby generator circuit breaker and restarting motors etc. This can be taken a stage further by having automatic changeover contacts between the mains and standby supplies and the switchboards on plant served, reducing manual operation to restarting motors etc.

c) No-Break System—this is the most expensive in both capital and running costs and is only used in extreme cases of individual plant items, such as a computer, where any break in electricity supply can result in very high losses. The alternator of the standby generating set is continuously run at operating speed by a mains driven electric motor. On mains failure, the standby generator prime mover is automatically started and the alternator connected to the supply via a changeover contactor.

A more recent system ideally suited to computers is a standby supply comprising basically a storage battery with an invertor floated across the load in parallel with the mains supply (Diagram 39). On the occurrance of a mains failure, the entire load is automatically taken by the battery/invertor, so maintaining an uninterrupted supply to the load.

The effective duration of operation of such an arrangement depends on the capacity of the battery in relation to the load, and should it be required to maintain the supply for more than an hour or two, it would be necessary to provide suitable backup by say a diesel generating set.

This system gives the very close control of voltage and

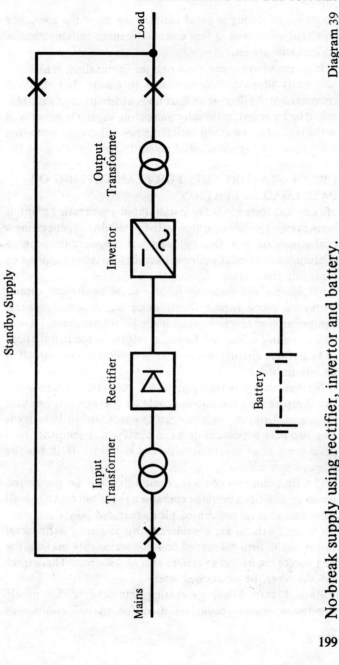

Diagram 39

No-break supply using rectifier, invertor and battery.

frequency necessary to avoid malfunctioning of the computer on changeover, and is less costly and more reliable than a no-break diesel set.

In cases where more than one set is installed, whichever scheme is adopted, it is necessary to ensure that the load connected to the first set on start up is within its capacity. This might be arranged by initially connecting say the lighting load to the first set and waiting until the other sets are synchronised before connecting substantial power loads.

USE OF STANDBY SETS FOR PEAK-LOPPING OR BASE LOAD OPERATION

If a case can be made out to install standby generating plant, it is always worthwhile considering the possibility of generating a proportion of the factory electrical requirements on a continuous or regular periodic basis before selecting plant or designing the system.

While this will certainly involve some additional capital outlay on more robust plant, more sophisticated control equipment and revenue expenditure on running costs; it will give a positive calculable financial return on the installation. In favourable circumstances, a good economic case can often be established.

No industrial generating plant without waste heat recovery can compete with the public supply on average cost per unit generated. However, most electricity purchased by industry is on a two part maximum demand tariff where the purchased price per unit of maximum demand is 100 to 200 times the average unit cost.

If a firm can reduce its maximum demand for purchased power by judicious private generation, there can be an overall reduction in total purchased plus generated power costs.

On the basis so far considered, this is purely a financial saving to the firm concerned and not necessarily making the best uses of the nation's primary energy resources. This aspect will, however, be mentioned later.

Basic Plant:- While generating sets selected for purely standby operation could be used for partial continuous generation, it will generally be more economic to consider more efficient and more robust units and a more permanent installation to minimise operating and maintenance costs. In the case of diesel engines, this could mean selecting lower speed engines and if the size of the units warrants it, using lower cost heavy fuel oil instead of diesel oil. Also if the size of the sets is beyond the range covered by radiator cooling of jacket water, a cooling circuit with cooling towers should be considered instead of a "once through" cooling system.

A decision will also have to be taken on economic grounds whether to operate in parallel with the public supply or supply only a segregated section of the works requirements from the private plant. This will largely determine the nature and cost of the electrical installation involved.

If parallel operation is opted for, particularly if peak lopping is the selected method of operation, it will generally be worth considering the installation of an additional generating unit to minimise the risk of savings being nullified by failure of the private plant.

It is these additional capital costs, over those of purely standby installation, which have to be weighed with running costs against the financial savings on the total works electricity costs.

Methods of Operation

If the works daily, weekly or monthly purchased electrical load curve has one or more substantial peaks, operating the private generating plant to supply these peaks may be the most economic method of operation.

If the load curve is relatively level, base load operation of the private plant may be indicated.

If peaks in demand are due to a small number of individual plant items, it may be possible to supply those plant items from the home generating plant on an electrical distribution system separated from the public supply. Base load operation can be practiced by operation in parallel with the public supply or sectioned operation of part of the total load.

Only a detailed analysis of the particular works load pattern

201

and an economic evaluation of all the possible alternatives, taking into account running costs and possible standby charges, will establish which method of operation will give the greatest financial savings for comparison with the additional capital costs discussed previously.

Waste Heat Recovery

Any scheme for partial power generation can be enhanced if it is possible to recover waste heat in the form of steam, hot water or hot gases for use in process.

If the potential for using such waste heat exists, there may be economic grounds for considering more expensive installations, such as higher pressure boiler plant and steam turbo alternators or gas turbo alternators.

Such combined heat and power installations have the added advantage of generally making more efficient use of the nation's primary energy resources than purchasing electricity and generating the heat requirements independently. Roughly 90% of privately generated electricity is produced on combined heat and power plants.

RELIABILITY

The reliability of any private generation scheme will naturally be compared with that of its alternative—the public electricity supply.

On average, the likelihood of interruption of the public supply to industry is appreciably less than two hours per year. As a percentage, the reliability of the supply is in excess of 99.98%. The very high standard of dependability is achieved by the installation of additional generation, transmission and distribution equipment and excellent standards of maintenance and fault recification. This is, of course, reflected in the purchase price of electricity produced.

The restrictions associated with the three day working week in early 1974 and the shorter crises of the two previous years have brought home to every one the vulnerability of the public supply to political and industrial action.

In the current social climate and world energy situation,

industry is reconsidering the protection of the means of production afforded by private generation. Failure of the public supply totalling one working week per year, for example, could reduce the reliability to below 98%.

The comparable reliability of private generating plant units depends on the extent of non-availability for generation purposes due to two factors—the time required for essential planned maintenance and overhaul involving shutdown and unplanned outage due to electrical or mechanical failure in service.

Some figures on programmed shutdowns can be obtained from manufacturer's and users records. Failures in service are unlikely to occur every year and can be so varied that realistic estimates are difficult to obtain. Reliability, taking account of these two factors, can be calculated assuming that the average industrial firm will have a holiday shutdown of two weeks per year during which planned overhauls can be carried out.

For typical prime movers the approximate indications are:-

Type of Unit	Programmed Shutdown	Failure	Overall Reliability
Diesel Engine	250 to 680 hrs. p.a. This could comprise one shutdown for major overhaul lasting two to three weeks and three to four weekend shutdowns.	94.0%	200 h. p.a.
Gas Turbine	Varies with type and manufacturer. 4 to 24 hrs. in first year of operation. 60 to 80 hrs. every second year. Four weeks every 16 years when set is virtually rebuilt.	98.7%	0-24 h. p.a.
Steam Turbine (excluding associated boiler plant)	72-180 h. p.a. depending on size. Possibly reblading 2 weeks every 10 years.	99.7%	0-24 h. p.a.

These figures refer to a single unit of each type. Obviously, the effect of a failure in operation can be reduced by having more than one unit which together can meet the maximum total load. Also, the installation of standby units in excess of total required rating will improve overall reliability and can enable maintenance or repairs to be carried out outside the annual shutdown period. The overall reliability of multi-set installations is a simple probability calculation.

When private generating plant is installed purely for standby duty to the public supply, the chances of a period of non-availability coinciding with a public supply failure are remote. A single unit of the required rating is all that need be installed, unless the works switching arrangement dictates the need for multiple units.

When private generating plant is used for partial generation on a reasonably continuous basis, either for peak lopping or as based load plant, the financial loss resulting from the failure of a single unit could be substantial. If diesel engines are used, it is usual to install at least two units to provide the required rating and depending on the financial implications of a particular case, it is occasionally economic to have a standby unit. If gas turbines or steam turbines are used, due to their greater reliability and higher capital cost, it is seldom economic to install more than a single unit for small to medium sized installations.

If private generating plant is used for total works power generation, the choice will tend to be the more reliable gas or steam turbines with full waste heat recovery. Some measure of standby is desirable, but the high capital cost of this is difficult to justify. One compromise which is frequently adopted, is to replace the plant before the end of its useful life and retain the older plant as standby. If space is available, this can be more economic than retaining a public supply as standby.

Fuel Supplies

A third major factor in considering the reliability of private generating plant is the availability of fuel. For purely standby generating plant the storage requirements of fuel can be assessed on the basis of the duration of public supply failure against which it is intended to insure, plus a reasonable margin for periodic operation of the plant to keep it in good condition.

The more continuously it is intended to operate the generating plant, the more important does it become to maintain adequate stocks of fuel to cover for possible breaks in supply.

Natural Gas

With natural gas on a continuous supply tariff, reliability is possibly on par with the public electricity supply with greatly reduced risk of failure due to political or militant labour action. British Gas have to be careful not to sell gas in excess of their supply capacity. The greatest risk of failure is probably accidental damage of local distribution mains. In these circumstances, it is almost certainly uneconomic to consider the installation of storage capacity.

With natural gas on an interruptible tariff, a secondary fuel, such as gas oil, is generally used to supply dual fuel plant or essential services in the event of interruption of supplies. Although interruptions of the natural gas supplies in the past have been local and few, and certainly nothing approaching the duration of breaks allowed for in the supply agreement, this situation could deteriorate in the next year or two at least until assured further supplies are available from the North Sea. It would be prudent to maintain stocks of the secondary fuel sufficient to cover the full periods of breaks allowed in the supply agreement. It must be remembered that in event of a supply shortage natural gas customers with only a standby requirement for oil will come at the bottom of the oil companies supply priorities.

In the past, oil companies have recommended a minimum oil storage of three weeks supply at maximum usage rate or two weeks at maximum usage rate plus one normal delivery, whichever is the greater.

Due to the past dependability of deliveries, many companies have installed storage capacity considerably less than this. If power generation is to be considered, it is recommended that three weeks remaining supplies at maximum usuage rate at any time is the minimum storage which should be considered.

Many companies are in fact now looking into doubling their past oil storage capacity to cater for possible prolonged interruptions in supply of up to two months. With heated grades of oil, as far as is possible, only one storage tank should be kept up to normal storage temperature and it should be thermally insulated to minimise heat losses. Similar storage is desirable for L.P.G., but the high capital cost of storage equipment may justify some compromise.

Coal

Unlike oil, coal deteriorates slightly in stock and this loss, and the additional handling costs, should be budgeted for. A minimum stock pile of around two monthly supply at maximum usage rate is not considered unreasonable. Many firms stock three months or more supply when space is available.

SUMMARY

Since the individual conditions vary so vastly between firms, no conclusive generalisations are possible. The following pointers, however, frequently apply.

1. Generation for standby purposes can give no guaranteed return on the capital expenditure involved and can only be justified on an insurance type calculation. As a result, the simplest possible installation with lowest post equipment consistent with reliability is first choice. In most cases this will be high speed diesel generators.

2. Peak lopping or base load partial power generation can only be justified on its own merits in very few cases. A sound economic case can often be presented, however, if:-

   a) It has already been decided to install standby generating equipment and only the addition capital cost over standby equipment has to be considered.
   and/or

   b) If a substantial proportion of the waste heat from the prime movers can be usefully recovered for steam raising, water heating or other process applications.

   c) If generation can be effected by back pressure steam turbines or by utilising steam or heat from process which would otherwise be wasted.

3. Total generation can result in the greatest financial savings, but is only applicable in comparatively few firms where there is a relatively large steam or heat requirement such that the power generation potential is in reasonable balance with or in excess of works power requirements. Captial costs tend to be high, but in favourable cases, so is the return. With steam turbine installations, the capital cost can be minimised if consideration of power generation is made when process steam boilers are due for replacement.

Any potential scheme for power generation requires the fullest evaluation of all the many factors in a particular firm and possible alternatives involved. Such evaluations should take full account of tax on savings, tax allowances, grants etc., and should preferably be evaluated on a discounted cash flow basis to establish a realistic payback period and the return on capital expenditure.

In the past, several schemes have been installed which are not financially viable and many schemes have been rejected because an insufficiently detailed analysis has been carried out to establish the true facts. Also when combined Power-heat schemes are bing reviewed the effect of possible future changes in process steam pressure or heat demand should always be carefully considered.

People tend to think of generation as involving electricity. It need not; indeed, some excellent small schemes use steam turbines to drive air compressors, pumps, fans, etc., where these operate continuously throughout the working period and where the exhaust steam can always be absorbed for feed water heating, process or space heating. Such schemes avoid all synchronising, sectioning or Electricity Board problems but effectively save the running cost of the electric motor that would otherwise have provided the power.

8
Heat Distribution

INTRODUCTION

Heat is essential to industry both for space heating and for a wide variety of process applications. In most cases it is not convenient, practical or economic to generate the heat at the point of usage, so it is necessary to provide a method of distributing heat which has been generated at a central point.

Water is one of the most readily available commodities known to man. One of the great legacies provided by nature to the engineer is that water, and its associated vapour form steam, both have physical properties which make them ideal media for the distribution of heat in the temperature ranges most commonly required in industry.

Steam and hot water should not be considered as raw materials, but rather as conveyors. If we were considering any other form of conveyor, such as a rubber belt or similar, we would be careful to ensure that it followed the most logical route, and that no material fell off, or was lost, due to bad design. Similarly with our steam and hot water systems we must ensure that they are designed so as to carry the heat safely, and with as little loss as practicable, to the various points of use.

CHARACTERISTICS OF WATER AND STEAM

In order to understand the value of water and steam as conveyors of heat it is necessary to learn something of their physical characteristics.

Sensible and Latent Heat

Upon the application of heat to water, the molecules of which the water is composed are caused to agitate at a speed increasing with temperature. When the temperature reaches "boiling point", the agitation has become sufficiently violent for the molecules to escape by breaking the surface boundary between the water and the air above. This stage marks the beginning of vaporisation. The heat required to reach this point from a base of water at freezing point is called the sensible heat of water, and is symbolised by the small letter h. It is called sensible heat because when we heat most things, the temperature rises i.e. it makes sense!

If the pressure above the water is raised, then a higher rate of agitation of the molecules, and thus a higher tempera-

211

ture, must be reached before the molecules can break the surface. This temperature is variously known as the boiling point, the saturation temperature, or evaporation temperature, and the graphical relationship between it and the relative pressure is shown in Diagram 2.

The relationship between the temperature of water and the heat applied to it, starting from water at freezing temperature, is shown graphically in Dia. 1. This shows that the temperature gradually increases from the freezing point t_0, in direct proportion to the heat applied, to a value t_1 shown as point A on the diagram, the quantity of heat being represented by h.

THE APPLICATION OF HEAT TO WATER

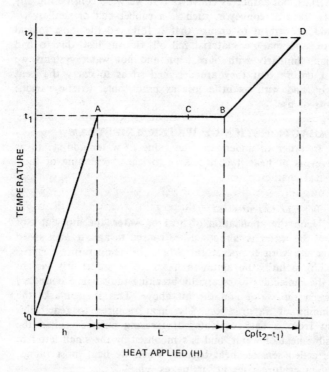

HEAT APPLIED (H)

Dia. 1

From this point onwards the further application of heat causes no temperature increase until all the water has been transformed to a vapour shown as point B on the diagram. At this point the steam is said to be dry saturated, but the temperature will be precisely the same, t_1, as that at which vaporisation began. The quantity of heat absorbed during this stage is known as the latent heat of vaporisation, and is denoted by the symbol L, in this cause it is called latent heat because it appears to be hidden i.e. there is no temperature rise!

Now the total heat, H, absorbed by the steam at its dry saturated state is the sum of the latent heat and the sensible heat and can be shown as

$$H = h + L$$

Dryness Fraction

At any intermediate point between A and B in Diagram 1 the extent of vaporisation can be measured as a fraction, which is termed the dryness fraction, frequently symbolised as q. This is the ratio of the latent heat absorbed at the intermediate point in question to the latent heat required to reach the point of dry saturation. If the intermediate point is C, as shown in the diagram, then the dryness fraction q will be $\dfrac{AC}{AB}$.

At C, or any other point along the line AB the steam will contain a certain weight of water, and this weight provides another way of identifying the dryness fraction. If W is the weight of a given quantity of wet steam, and w is the weight of water suspended in it, then the dryness fraction, $q = \dfrac{W - w}{W}$

Although the dryness fraction of steam is simple to define, as indicated above, it is extremely difficult to measure with any degree of accuracy, particularly at values which approach the dry saturated condition. The main difficulty occurs in obtaining a sample which is genuinely representative of the average condition of the steam in the pipes.

We can now modify our equation for the total heat of steam up to the dry saturated point as follows:

$$H = h + qL$$

213

This relationship can be transposed to show the dryness fraction in heat terms as:

$$q = \frac{H - h}{L}$$

which is the same ratio as that given by the ratio AC in Dia. 1.

$$\frac{AC}{AB}$$

Superheat

If more heat is added to the dry saturated steam after point B on the diagram, then the steam is said to be super-heated, and the level of superheat is measured by the increase in temperature which takes place. Provided that the steam is allowed to expand freely then the pressure will remain constant with the addition of superheat, but the density will reduce. Since the application of this additional heat results in a rise in temperature it is once again 'sensible' heat.

Referring again to Diagram 1, if D is the point ultimately reached by the steam, at which the temperature is t_2, then the extent of superheat is measured as $t_2 - t_1$. The quantity of heat required to reach this point is given by Cp $(t_2 - t_1)$ where Cp is the figure for the specific heat of steam at constant pressure. The value of Cp is generally taken as 0.48, but this can only be a broad approximation since the actual figure varies with both pressure and temperature, increasing with the former and decreasing with the latter. It is thus necessary to refer to the Steam Tables to obtain an accurate figure for any given condition.

The total heat of steam superheated to a temperature of t_2, and reckoned above a datum of freezing point, can now be given as:

$$H = h + L + Cp \, (t_2 - t_1)$$

Steam Volume

Steam is continually increasing in volume during both the vapourising and the superheating stages. In fact the volume of steam in all stages varies with both temperature and pressure, andcan be obtained for dry saturated steam and for super-heated steam by reference to the Steam Tables.

The volume (V) of wet steam of a given dryness fraction (q) can be simply calculated from the volume (V_s) of dry

214

steam as follows:
$$V = q \, Vs$$
A useful approximation (accurate to $\pm \frac{1}{2}\%$) for the volume of superheated steam is given by the formula
$$V = \frac{1.253 \, (H - 835)}{P}$$
where V is the required volume in cu ft/lb
H is the total heat in Btu/lb
P is the pressure in psi absolute

Steam Density

The density of steam is frequently required in simple calculations, and is given, of course, by the reciprocal of the specific volume
$$\text{ie Steam Density } w = \frac{1}{V}$$

Steam and Water Tables

The variations in physical characteristics of steam and water are so wide that their numerical values can only be readily dealt with in graphical or tabular form, and the Steam and Water Tables are the most convenient form for most reference purposes. With these tables it is possible to read directly, or to calculate, whatever property of steam or water may be required. Heat contents are normally given as those in excess of water at freezing temperature, as a standard basis. Detailed tables can be purchased in book form suitable for the student, design engineer etc but abbreviated forms are given in many technical diaries and handbooks in sufficient detail for the practical engineer.

HEAT DISTRIBUTION USING STEAM

Choice of Pressure

It has been said that, ideally, a steam pipe should carry steam by the shortest route in the smallest pipe with the least heat loss and the smallest pressure drop that the circumstances will permit. The higher the pressure the more steam a given pipe can carry for the same pressure drop. However, the higher the pressure, the higher the temperature, and therefore the

215

higher the heat loss. It is obvious that the choise of pressure must involve some compromise.

For most processes the greatest economy will be achieved by using steam at the lowest practicable pressure. On the other hand boilers, and particularly modern boilers, are unlikely to operate satisfactorily at pressures below their design condition.

In most small to medium plants the plant requiring the highest steam pressure frequently sets the main distribution pressure, and in many cases this sets the operating pressure for the boiler plant.

As a starting point it is well worthwhile to experiment with process plant to ascertain the lowest pressure that will give satisfactory performance. This can be readily done by fitting a pressure gauge on each item after the inlet valve and gradually reducing pressure until a fall-off in performance occurs. When this information is available on all items of process plant a start can be made on assessing the optimum steam distribution pressure.

Steam Pressure v Steam Volume

The prime factor in establishing the carrying capacity of a steam main is the maximum permissable velocity, and thus, the quantity in volumetric terms which can be carried. The higher the pressure of the steam the lower the volume, and thus the lower the pipe size required to provide a velocity within the acceptable limits.

The relationship between the pressure and volume of saturated steam is shown in Diagram 2 for the pressures in common use. The rate at which the volume increases with pressure at the lower ranges presents the strongest argument in favour of distributing steam at high pressure and undertaking any pressure reduction local to the process.

Use of Reducing Valves

A system frequently used is to operate the boiler plant at its design pressure and fit a main reducing valve at the start of the distribution system. The mains are then sized generously so as to give a small pressure drop, the main reducing valve being set as required by the plant needing the highest pressure. This system has some advantages, since at peak loads the boiler pressure can be permitted to fall to the

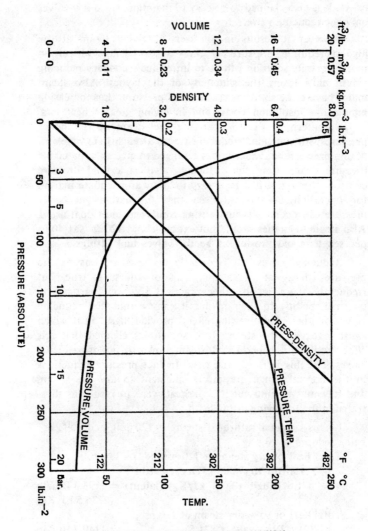

THE VARIATION WITH PRESSURE OF THE PHYSICAL
CHARACTERISTICS OF DRY SATURATED STEAM
Dia. 2

217

'reduced' steam pressure, thus providing some 'accumulator' effect. It is also claimed to be easy to maintain, since it involves only one reducing valve.

However it results in all users receiving steam at the highest required pressure. If this causes wastage at certain users, the only solution will be to introduce secondary reducing valves, thus losing the simplicity of the layout. Also steam mains have to be sized for the reduced pressure, thus increasing piping and insulation costs, and involving greater heat loss.

An alternative is to distribute the steam at full boiler pressure with individual reducing valves at each item of plant. Mains sizes can be reduced, and arranged to give an appreciable pressure drop, using the reducing valve as a final pressure control. In practice it is necessary to allow an adequate margin for operation of the valve so that fluctuations in boiler pressure can occur without causing trouble at individual users. Also steam velocities should not exceed about 120 ft (say 40m) per sec to avoid erosion at bends, valves and fittings.

Reducing valves are frequently criticised as wasting the pressure energy of the steam. This of course is true, but frequently there is no other practical solution, particularly on small plants. In their favour it can be said that reducing valves, in their wire-drawing effect, provide 'flash' heat which assist in drying the steam. It is sometimes claimed that this effect results in superheated steam downstream of reducing valves, but this is rarely the case. In fact pressure reductions in the normal range are rarely sufficient to dry the steam, and it would be unusual if superheat were to result, as illustrated in the following example.

Let us consider saturated steam at 120 psig and 95% dry being reduced to 20 psig

Total heat of dry steam at 120 psig (827.4 kN/m^2)

$= 321.8$ Btu/lb (748.5 kJ/Kg) (sensible)
$+ 871.5$ Btu/lb (2,027 kJ/Kg) (latent) $= 1193.3$ Btu/lb
(2775.5 kJ/Kg)

Total heat of 95% dry steam at 120 psig

$= 321.8$ Btu $+ \dfrac{(95 \times 871.5)}{100}$ $= 1149.7$ Btu/lb
(2674.2 kJ/Kg)

Total heat of dry steam at 20 psig $= 1167.6$ Btu/lb
(2715.8 kJ/Kg)

Deficit of heat available to dry the steam = 17.9 Btu/lb
(41.6 kJ/Kg)
Latent heat of steam at 20 psig = 940.1 Btu/lb
(2186.7 kJ/Kg)
Therefore extent of wetness remaining in
LP steam = 17.9 x 100 = 2% approx

940.1

ie A reduction from 120 psig of steam which is 95% dry will give steam at 20 psig and 98% dry. In fact reference to the steam tables shows that only by a reduction to atmospheric pressure would the original steam have become completely dry.

Alternative Methods of Pressure Reduction

Before resorting to the use of reducing valves all other practicable means of making better use of the energy in the steam should be considered.

ENGINES AND TURBINES. The ideal solution is to generate steam at sufficiently high temperature and pressure so that it can be used to provide motive power via an engine or turbine. The steam is ultimately discharged at an intermediate (pass out) and/or final (back pressure) stage at conditions arranged to be suitable for subsequent process or space heating use.

Various aspects of this have been dealt with in a preceding section and where suitable conditions exist it presents the most satisfactory way of utilising the total energy available in the steam. The motive power is usually considered primarily for possible electrical generation. It is unfortunately true however that the application of this principle is limited by the high initial cost of the plant involved, the cost of standby plant, and the relative costs of purchased elecericity. Nevertheless, wherever a suitable balance appears to exist between the steam and electrical requirements of a plant, the economics of on-site power generation should be studied in detail.

An alternative possibility not frequently considered, but which may well merit examination, is the use of a steam engine to drive an air compressor. This could result in a substantial saving in electricity consumption, with standby problems very much less than in the case of plant for generating electricity.

219

STEAM ACCUMULATORS. Accumulators can be extremely useful in situations where a boiler plant is required to meet a widely fluectuating load, and where steam is required at two or more well separated pressure conditions. The principle involved is that during 'troughs' in steam demand, the surplus steam being generated by the boiler plant is used to heat water contained in a pressure vessel. During 'peaks' of steam demand this hot water is allowed to 'flash' off in order to provide steam to the lower pressure processes. By careful and correct selection of the various operating conditions the boiler plant can be operated at relatively steady firing rates, despite the load fluctuations, and steam pressures can be maintained at a constant level.

EVAPORATORS. Pressure reduction can frequently be suitably undertaken by the use of an evaporator, to replace a reducing valve. Although in thermo-dynamic terms it is no more efficient than the reducing valve it can be arranged to produce a substantial quantity of clean condensate, effectively distilled water, as a by-product of the pressure reductcion.

Steam Pipe Sizing

In sizing the piping for a steam distribution system it is, of course, essential to know the complete load pattern throughout the cycle of operation on the system. A process using a given quantity of steam in an hour, may, at certain times during the hour be using steam at several times the average rate over the hour. It is obvious that the steam supply pipe must be sized to cater for the maximum rate of flow if satisfactory conditions are to be maintained throughout the cycle.

Arriving at the appropriate pipe size involves a compromise between velocity and the optimum pressure drop for the particular situation, together of course, with the limits of the standard pipe sizes available.

Recommended maximum velocities in pipes are:

	Feet/second	metres/second
Vapour under vacuum	150-200	45-60
Wet exhaust steam	70-100	20-30
Dry saturated steam	100-130	30-40
Superheated steam	150-200	45-60

The calculation from first principles of pressure drops for given conditions is complex, but the data can be readily obtained from nomograms and curves given in various reference books and manuals. The information is normally provided in terms of the pressure drop for unit length of straight pipe, under given steam conditions.

Valves, fittings, bends and elbows all add to the overall pressure drop and these are quoted in terms of an equivalent length of straight pipe. Thus the equivalent pipe lengths for all the fittings are added together with the actual pipe length involved, and the result multiplied by the pressure drop for unit length of the pipe to give the overall resistance of the system under consideration. The answer to this calculation may indicate that a larger pipe size, ie, a lower velocity, must be selected in order to reduce the pressure drop to a satisfactory level to ensure the required pressure at the user end.

It is of major importance to consider the actual steam pressure in the system when sizing a steam pipe, and not to use, say, the nominal boiler pressure. A modern boiler with automatic control may often modulate over a range of pressure for at least 15 psi (say kN/m²); thus if it is nominally working at 145 psig (1000 kN/m²) maximum, it will periodically drop at least to 130 psig (900 kN/m²). The peak steam requirement will often coincide with minimum boiler pressure, and if there is any danger of creating a short-term overload on the boiler then the pressure may drop further befofre it starts to recover. Our calculation must be based on the mean pressure in the system, and not just on the system inlet pressure, so we must make an allowance for the pressure drop through the system. An example will illustrate some of the points made.

What is the maximum pressure which we can reasonably expect to achieve at a process plant requiring a maximum flow rate of 4,500 lbs of steam per hour (570 Kg/s), through a 3″ main from a boiler with a nominal working pressure of 145 psig (1000 kN/m²)?

Measured length of mains, say	300 ft
	(915 m)
Equivalent allowance, from tables, for valves and fittings, say	200 ft
	(610 m)
Total equivalent length of mains	500 ft
	(1525 m)

221

Nominal boiler pressure	145 psig
	(1000 kN/m^2)
Allowance for modulation range	15 psig
	(100 kN/m^2)
Maximum reliable pressure at inlet to system therefore	130 psig
	$(900 \text{ kN/m}^2 \text{ approx})$
Allowance for mean pressure drop in system say	5 psi
	(35 kN/m^2)
Mean pressure in system therefore	125 psig
	$(865 \text{ kN/m}^2 \text{ approx})$

Reference to nomograms on pipe sizes, pressure drops and velocities gives the following information:

Velocity of steam in 3″ pipe carrying 4,500 lbs of steam per hour at a mean pressure of 125 psig	85 ft per sec approx
	(26 m/s)
Pressure drop per 100 ft of pipe	2 psi
	$(14 \text{ kN/m}^2 \text{ approx})$
Therefore total pressure drop through system	10 psi approx
	$(70 \text{ kN/m}^2 \text{ approx})$
Therefore pressure available at user	120 psig
	(830 kN/m^2)

The nomograms have indicated that our steam velocity at 85 ft per sec is well within the recommended maximum, and have confirmed that our original allowance of 5 psi as the mean pressure drop is about right.

Nevertheless we have to conclude that the maximum pressure we can rely upon at the user is only 120 psig, although we started with a boiler nominally rated to produce steam at 145 psig.

It is obvious that the available pressure at the user will be again reduced should our boiler modulating range be greater than the 15 psi allowed in the calculation, or should our peak load cause any additional pressure drop on the boiler. This effect will be increased by the greater pressure drop arising from the consequently lower mean steam pressure in the system.

222

Water in Steam and Drainage

It has been stated that saturated steam is rarely, if ever, completely dry and the average wetness in a system is likely to be 5% or more. This wetness is a source of embarrassment since the water is of no practical use, and will ultimately reduce the rates of heat transfer in the process plant. It is carried as droplets in the stream, which dependant upon the velocity of the steam can be highly erosive at bends and fittings. It provides additional loading on trapping and condensate systems and creates additional 'flash' when released to atmospheric pressure.

Furthermore, the collection of water in steam mains can produce extremely dangerous stressing due to water-hammer.

It is imperative that water in the steam should be removed in order to present as dry a steam as possible to the process. Steam driers, or separators, are frequently installed which generally use a centrifuge principle to throw the water out to a collection point. A steam trap is used to remove the separated water. The separator should be placed as close as practicable to the point at which the dry steam is required.

Since some water is inevitable it is necessary to provide drain points fitted with steam traps at all low points of the steam system. All steam mains should preferably be arranged with a slight fall in the direction in which the steam is flowing.

It should be noted that where the natural drainage of a steam main is against the steam flow then the pipe sizing should be increased to allow for the disturbance to conditions caused by the contra-flow of the steam and water.

Ideal drain pockets are equal T pieces to give a large collecting area. Two typical arrangements are shown in diagram 3 below.

Where heavy initial condensation is likely a piece of pipe can be fitted to the T to give larger storage volume, so enabling the trap to discharge the surge over a longer period of time. A useful precaution is to take the pipe to the trap from a few inches above the bottom of the pocket, to allow scale and sediment to settle. A drain valve or plug can be fitted for periodic cleaning out. All branches should be taken from the top of the main, otherwise they may collect condensate as well as steam.

223

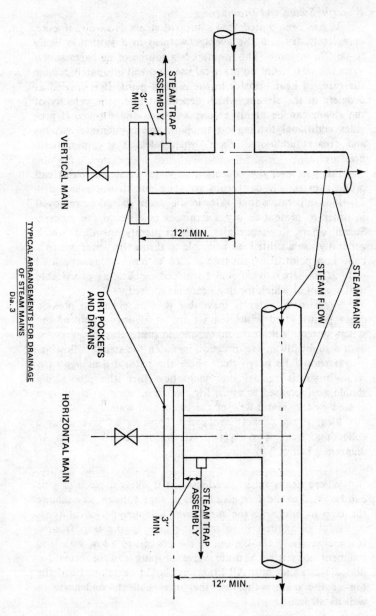

TYPICAL ARRANGEMENTS FOR DRAINAGE
OF STEAM MAINS
Dia. 3

VERTICAL MAIN

STEAM TRAP
ASSEMBLY

3"
MIN.

12" MIN.

DIRT POCKETS
AND DRAINS

STEAM FLOW

STEAM MAINS

HORIZONTAL MAIN

STEAM TRAP
ASSEMBLY

3"
MIN.

12" MIN.

224

Superheat in Steam Distribution

Where practicable there is often a case for feeding into the distribution system steam which has a small amount of superheat. It can be shown that superheated steam does not release its heat so readily to a metal surface, such as a pipe wall, as does saturated steam. Pipe losses tend therefore to be reduced, provided the superheat temperature is not excessive. The more important point is that the pipe heat losses are met by reducing the superheat temperature rather than by causing wetness in the steam. With the correct choice of superheat the process plant can be supplied with dry steam, thus improving heat transfer and reducing the danger of water logging. Mains drainage systems can be simplified and of lower capacity in view of the lower quantity of water being condensed in the main.

Selection of Traps for Steam Mains

There are many types of steam trap and selection of the correct trap for a given application is of major importance in order to ensure satisfactory performance, avoid losses and avoid undue maintenance problems. Traps for steam main drainage should ideally have the following characteristics.

: A high discharge capacity to meet start-up conditions.
: Capacity to discharge air.
: Ability to withstand water hammer.
: Must cope with pressure fluctuations.
: Preferably continuous discharge.
: In external applications must be well insulated, or of a type that does not retain water.
: Superheated steam applications require special consideration.

Trap manufacturers invariably provide data on the characteristics of the various types and a selection must be made to meet the above criteria with as little compromise as possible. The size of trap selected should be related to the quantity of condensate being discharged and not necessarily to the pipe size involved.

It is important that each trap should be equipped with the required ancillary fittings and the ideal arrangement is illustrated in diagram 4.

225

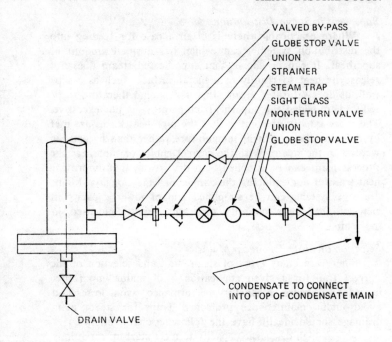

VALVED BY-PASS
GLOBE STOP VALVE
UNION
STRAINER
STEAM TRAP
SIGHT GLASS
NON-RETURN VALVE
UNION
GLOBE STOP VALVE

CONDENSATE TO CONNECT
INTO TOP OF CONDENSATE MAIN

DRAIN VALVE

Dia. 4
EQUIPMENT IN STEAM TRAP ARRANGEMENT

Condensate and Flash Steam Usage

Condensate formed under pressure contains an excess of sensible heat which it cannot normally retain when the pressure on it is reduced. On pressure reduction the excess heat above the new lower sensible heat figure causes a proportion of the condensate to 'flash' into vapour. Unless proper arrangements are made for collecting this flash it appears as an unsightly billow of vapour, carrying away heat and pure water to waste, adding to the cost of water treatment etc. Taking the case of condensate formed from steam at 100 psig (690 kN/m²) and discharged at atmospheric pressure, some 12% of the original condensate is inevitably lost in this way.

With properly designed equipment this surplus heat can be collected and used in a lower pressure system. Again assuming steam at 100 psig the following calculation illustrates the

226

amount of steam which can be made available, assuming a low pressure system operatingg at 5 psig (35 kN/m²)

Sensible heat of condensate at 100 psig	309 Btu/lb (719 kJ/Kg)
Sensible heat of condensate at 5 psig	195.5 Btu/lb (455 kJ/Kg)
Therefore heat available on pressure reduction	113.5 Btu/lb (264 kJ/Kg)
Latent heat of steam at 5 psig	960.8 Btu/lb (2235 kJ/Kg)
Flash steam formed per lb of condensate = $\dfrac{113.5}{960.8}$ =	0.118 lbs

Thus 11.8% of the condensate collected will provide low pressure steam at no cost other than the capital cost of the flash vessel and associated fittings. Put in another way, one pound of low pressure steam can be obtained without any additional fuel cost, for every nine pounds of high pressure steam used, and the condensate will still be available, at the more convenient temperature of the low pressure steam system.

To quote one factory engineer, no doubt much over-worked, "If only steam would vanish to nothing when it condensed, it would save a lot of trouble running pipes to the nearest drain." This is not quite the correct attitude, as the ideal place to put condensate is the boiler feed tank. Here it provides economies in three different ways, viz:

It provides heat with which to preheat boiler feed water.

It saves the cost of the raw water which would have been necessary to replace it as make-up.

It is of adequate quality for boiler feed thus saving raw water treatment costs.

A frequent difficulty is that the condensate returned is so hot that the feed water becomes too hot, particularly where high pressure condensate is concerned, and/or where a high percentage of condensate return can be obtained. We have already illustrated how high pressure condensate can be flashed off into low pressure steam. This steam can be used in the normal conventional way, through low pressure processes and

227

heating systems, giving condensate at a more practical temperature.

Where a flash system cannot be used it is often possible to devise other valuable uses for the heat in the condensate. If the steam is being used for space heating equipment ie radiators or unit heaters, the condensate can be collected from a suitable number of units and itself passed through an additional unit to extract heat before it is returned to the main condensate system. In driers, bottle washers etc it is often possible to use the condensate through a coil in a preheating section of the unit. It is not possible to generalise but a little ingenuity can usually illustrate a way in which the surplus heat can be used. It is important that the condensate should be properly dealt with at the boiler feed tank, ie taken below the water level and discharged either through a sparge pipe or via a heating coil.

On almost every boiler feed tank seen by the author, the condensate return pipes end well above the water level, so that any steam is usually lost to the atmosphere. The usual argument is that if the return lines dip below the water surface, water is sucked back and fills the return lines and even the steam spaces in the plant once steam is shut off and the plant cools down. However, a "vacuum breaker" could be fitted, comprising either a flap type non-return valve, mounted so as to admit air if any vacuum starts to form, or a small jet mounted in the return pipe above the water level. With the jet, a very small negligible amount of "flash" steam emerges from it while condensate is flowing, but air enters and prevents any syphoning action as soon as the steam is shut off at the plant concerned.

A small hole drilled in the pipe, above the water level, would be just as effective, but such a hole tends to enlarge at a surprising rate, and it is much better to use a brass jet (gas type) with a small drilled hole of up to $\frac{1}{8}$ in. dia., since this lasts much longer and can easily be replaced when worn.

In many cases, condensate mains are over-sized at the design stage, usually to make certain that little back pressure exists at traps. This often causes the scheme to be abandoned for capital cost reasons. Whatever the faults of steam traps of a century ago, modern ones are robust and of a relatively simple design and can stand back pressure on condensate lines providing they are installed correctly and receive suitable maintenance. It must be appreciated that almost every design of trap on the market today discharges condensate at the pressure of the steam in the plant the trap is draining, apart from a slight pressure loss due to the velocity of the water passing through the trap. Consequently, for every one pound of steam pressure before the trap, the condensate can safely be lifted two feet, or can overcome a resistance to flow of two feet of water friction head loss. Tables are available in technical literature giving such friction losses in steel or copper pipes, and providing allowances are made for the additional friction losses through valves, bends and fittings, they can be used to size condensate return systems. The pipe diameter thus obtained is often far smaller than "rule of thumb" methods suggested, and may so lower the capital cost as to make a scheme attractive. It is, of course, important to take the lowest steam pressure, if condensate from a plant at several steam pressures is being collected.

Air Venting

The removal of air, and other foreign gases such as CO_2, from steam is of vital importance to fuel economy but is perhapps the problem which receives least attention. Its main disadvantages are as follows:

TEMPERATURE REDUCTION. In a mixture of air and steam they will each contribute to the total indicated pressure approximately in proportion to their relative volumes. The steam temperature is consequently lowered to that corresponding to the pressure which it is exerting.

HEAT CAPACITY. The capacity of the mains to convey heat is reduced, both by the lower temperature arising from its reduced effective pressure, and by the reduction in steam carried due to the presence of the air.

REDUCED HEAT TRANSFER. The air forms a film, which is highly resistant to heat transfer, on the heating surfaces of the steam user.

229

CORROSION. The oxygen in the air carried with the steam, and the CO_2 frequently present, can cause corrosion problems throughout the entire steam and condensate system.

With any main subject to intermitten use it is impossible to exclude air. On shut down a vacuum is formed which inevitably draws air in at glands and every other possible entry point, so that on startup the piping system is full of air which must be removed.

Air venting traps have already been mentioned and these, where suitable from other viewpoints, perform an invaluable function. Ideally additional means of air venting should be provided at all high and/or remote points, either by automatic air vents or by manual venting valves.

HEAT DISTRIBUTION USING HOT WATER
Low Temperature Hot Water

These systems convey heat by means of water at temperatures up to atmospheric boiling point. Modern pumped systems are normally designed on an overall temperature drop not exceeding about 11-12°C, equivalent to some 45 to 50 kJ per kg of water passing.

Systems are normally pressurised by the use of a head tank whose functions are:

: To ensure that the system remains full at all times.
: To absorb thermal expansion in the contents of the system.
: To maintain a slight pressure "head" on the system so that
: atmospheric boiling temperatures can be approached with-
: out danger of creating boiling conditions anywhere in the
: system.
: The normal upper tempeature limit on the boiler plant is
: about 90°C.

The head tank should be sited well above the highest pipe in the system and should be of adequate size to absorb the total expansion in volume from normal cold conditions to working temperature without over-flowing.

Where it is inconvenient or impracticable to provide a suitable siting for the head-tank then this can be replaced by a pressurising unit. This is essentially a sealed vessel with a membrane, having pressurised air on one side and water from

230

the system on the other. The air-side of the vessel is initially charged to a pressure equivalent to more than the height of the system, so as to maintain the system full of water.

With LTHW systems the energy transfer per unit of water circulated is relatively low, and for this reason pipework and pumps are relatively large. The low pressures involved make it easy to use, and the plant and controls can be unsophisticated and simple. It remains ideal for small heating systems in office blocks, commercial premises, etc.

Medium Temperature Hot Water

This term is usually applied to systems operating at slightly higher temperatures than LTHW but limited to the maximum conditions permissible with equipment manufactured to the same standards as for LTHW. The maximum flow temperature is normally 100°C to 120°C, i.e. sufficiently high to require that pressure must be applied, or generated, in order to avoid steam production.

These slightly higher temperatures enable higher heat transfer rates to be achieved from the heating equipment, thus permitting a reduction in pipe sizes. System design temperature drops are normally in the region of 15-16°C, equivalent to about 70 kJ per kg of water passing.

In systems at these conditions, head tanks alone cannot provide sufficient pressure to guard against boiling, and slightly more sophisticated pressurising systems are required.

Head tanks can be used for cold charging of the system via a non return valve. A membrane vessel can then be incorporated to provide the pressure and expansion space required as the system heats up. Expansion within the system creates a sufficient pressure in the membrane vessel to ensure that boiling does not occur at the temperatures in use, even in the highest parts of the system. A simple variation to this is where such a system is charged via pumps, rather than by a head tank.

An alternative which is sometimes used in the MTHW range is a pressurisation unit consisting of tank, pumps and pressure vessel. The pressure vessel is part filled with water with a pressurised inert gas above it. As heat is applied to the boiler so the water level rises in the pressure vessel and excess water is passed out via a relief valve to the water storage tank.

231

As the water level falls a pump operates to maintain the water level. The pressure of the inert gas is automatically maintained.

High Temperature Hot Water

Increasing the pressure applied to the system permits increased flow temperatures of the water from the heater. The upper limit is established by the pressure standards for the system, allowing for the reduction in static pressure at the highest point of the system or for the depression at the pump suction.

The higher flow temperature permits still greater improvement in heat transfer rates, and the higher permissible temperature difference provides for reduced quantities of water in circulation, with consequent reduction in sizing of plant and piping. Design temperature differentials are normally in the region of 35 to 40°C equivalent to some 150 to 160 kJ per kg of water passing. The practical limit is the difficulty of balancing the very small flows required by individual users at higher temperature differentials.

Originally HTHW systems used steam boiler plants, drawing hot water from below the normal operating water level. The boiler plant was controlled at a given static steam pressure. A proportion of the return water was mixed with the flow for temperature control purposes and a separate by-pass was included local to the boiler for attemperation, to ensure that the temperature throughout the system remained below the equivalent saturation temperature.

The above system, known as "steam cushioning", proved satisfactory on single boiler installations but difficulties were experienced in maintaining stable water levels with two or more boilers operating together. The control difficulties are such that this system is no longer favoured.

Modern HTHW systems generally use fully flooded boilers with separate pressurisation units similar to the method described for MTHW plants. Alternatively there are plants available which operate by injecting steam into hot water in a vessel where pressure is maintained by a stea mcushion.

Pipe Sizing for Hot Water Systems

It has already been illustrated that there are limiting temperature drops for each type of hot water system. Thus for a given total heat requirement the quantity of water which must be circulated can be readily calculated. With the circulating quantity fixed and the length of the circulating system known it is possible to choose parameters of velocity and pipe pressure loss to give the optimum related conditions of pipe size and total circuit pressure differential. Tables giving this data are readily available in design manuals and handbooks, and the method of calculation is similar for all types of hot water systems.

In hot water systems the relationship between water flow rates and pressure drops is vital and fundamental to the operation and balance of the system. A variation in flow rate in one section of the system will automatically alter the flow pattern of the remainder of the system.

The above comment illustrates an important difference between steam and hot water systems. Reasonable additions to steam systems can be made and provided there is sufficient steam available the addition will have, at worst, only a marginal affect on the remainder of the system. Additions to hot water systems however, cannot be made without a likelihood of upsetting the balance throughout the entire system. In certain cases additions made simply by connecting into adjacent mains have made the system impossible to balance. Additions can only be satisfactorily made by a complete reappraisal of the design of the system, followed by a rebalancing exercise.

The effect described above is most likely to arise in connection with HTHW systems in so far as these are normally extensive, with high flow rates and critical pressure differentials.

Imbalance in HTHW Systems

Imbalance is probably the most important single factor affecting fuel economy in hot water systems, and particularly HTHW systems. It can arise not only through additions to the system as described above, but through faulty initial design; using the wrong pattern valves for balancing purposes; the use of unbalanced diverter valves at remote control points; and/or the use of automatic throttling valves for process control purposes.

The results of imbalances have a progressive effect. The initial short circuiting which occurs causes the temperature differential to drop across flow and return. This, together with

233

reduced circulation at the extremities, results in cold spots. Normal response to this condition is to increase the on-line pumping capacity which may correct the low circulation at the extremities. However, the short circuiting will be tremendously increased with very high flow rates within the boiler house. Such high flow rates are frequently sufficient to cause a major restriction in the boiler output resulting in more boilers being on-line to meet a given load. Thus energy is being wasted by excessive pumping capacity, over-heating in certain localities and low overall efficiencies of boiler plants. In certain cases the overall effect has been so severe that a boiler plant has been unable to meet a normal winter heating load although it has an installed capacity equal to twice the requirement.

INSULATION OF DISTRIBUTION SYSTEMS

Faced with ever escalating fuel costs it is hardly necessary to stress the vital importance of insulating hot surfaces to reduce the loss of heat into the atmosphere. However the standards by which insulation is applied and maintained are frequently below what they should be and it is worthwhile considering some of the principles involved. The heat lost from a hot surface is a function of the following factors:

1) Temperature of the surface relative to its surroundings.
2) Air velocity over the surface.
3) Finish and colour of the surface.
4) Situation: internal, exposed or enclosed.

It can readily be deduced from the above that the heat loss will be higher the greater the temperature of the surface involved: that it will be higher in moving air than still air: that it will be higher in the open air than in an enclosed space. The selection of the actual insulating material for a given application is well covered in manufacturers' literature, and provided the material complies with the appropriate British Standard, the choice can be based purely on thermal properties and price. The main problem lies in selecting the optimum economic thickness.

The most important thing is that there should be some surface covering. The savings from pipe insulation are a good example of the law of diminishing returns. The first layer of insulation is the most important.

A coating of a metallic paint, ideally aluminium paint is very effective and will reduce heat loss by some 12 to 14%. This could obviously only be satisfactory on relatively low temperature surfaces, and in any case should not be used on surfaces of high enough temperature to cause it to discolour, say, 120°C. Although some engineers prefer BSI colour-coded pipes the use of low emmissity paint on the external surface of lagging can reduce surface losses. Alternatively polished aluminium cladding with occasional bands of colour for identification can be used.

Nomograms and tables to facilitate the calculation are available in many publications, technical handbooks and in manufacturers' literature. Some of the relevant factors have been dealt with earlier in this publication on pp. 3-26 to 28.

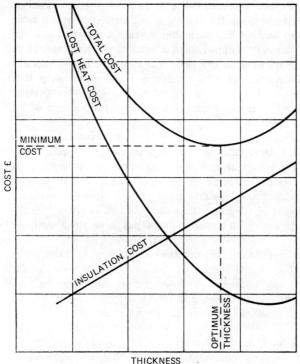

OPTIMUM THICKNESS FOR INSULATION
Dia. 5

235

A basis for the calculation is illustrated in the graph in Diagram 5. It is first necessary to establish the desirable pay back period for the insulation cost. The discounted costs over this period are then plotted for increasing insulation thickness. On the same ordinates a further curve is plotted of the cost of the heat lost over the period, which decreases with increasing insulation thickness. The summation of these two costs as a total cost will show a minimum point, indicating the economic thickness of lagging for the particular application.

The comments above have dealt with the application of new insulation. In most factories a critical appraisal of the condition and effectiveness of existing insulation would yield a very good return. It is frequently the case that, while pipes are adequately insulated, valves and flanges are left bare. A flange is at least equivalent to a foot run of pipe, and even on single shift operation the lagging of a flange on a steam main is likely to recover its cost within a year. A moulded box can be used, for easy removal, with a small drain pipe inserted to give early warning of any leak. Valves and other fittings can be similarly enclosed with insulating boxes. Very often it is found that the total surface area of flanges, valves and fittings, normally left bare, is a significant proportion of the area of the piping involved.

All insulation must be adequately protected against mechanical damage. Pipes at low level are often continually stepped on with consequent deterioration of the lagging. A simple wooden bridge across the pipe can save this damage. Where ladders are regularly leant against pipework a special metal sheath of adequate strength should be provided. Insulation of pipework in covered ducts has been dealt with on p. 3-28

Weatherproofing of external insulation is particularly important since the ingress of moisture seriously impairs the insulating characteristics. The weatherproofing must be continuous and impenetrable, and special attention is required to this where changes in section and direction occur.

Pipework for Heating Systems

The previous comments assumed that all pipework is insulated. It may be argued that this is not necessary for steam mains which *only* supply space heating equipment, if the mains

236

DIAGRAM 5A

	HEAT GIVEN OUT BY 1ft. (30cm) OF 1½in. (38mm) PIPE CARRYING STEAM AT 30lbf/in.² (2.1 Bar.) Btu/hr (Watts)		
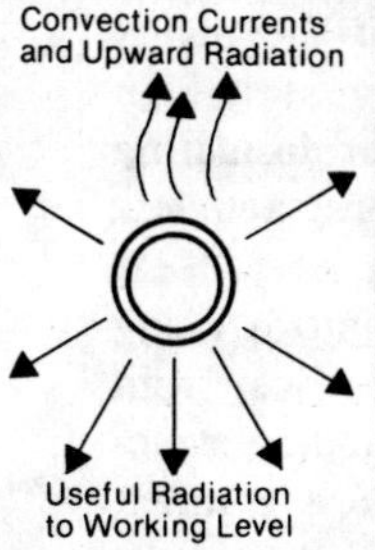 Convection Currents and Upward Radiation	Radiation Downwards	Losses Upwards	Total Heat Emission
	A. Bare Pipe		
Useful Radiation to Working Level	110 (33)	271 (81)	381 (114)
Insulation, Reflector, Joint	B. Converted to Radiant Strip		
	330 (99)	180 (54)	510 (153)

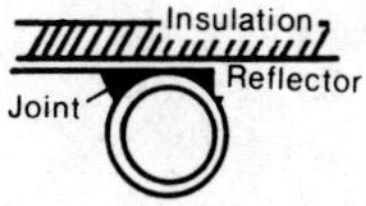

EFFECT OF CONVERTING OVERHEAD PIPE TO RADIANT STRIP

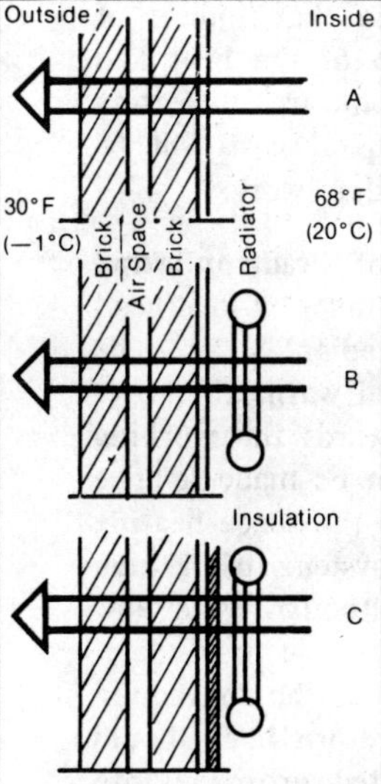

	Heat Losses Through Wall Btu/ft²/hr. (Watts/m²)	RADIATORS PLACED AGAINST OUTSIDE WALLS
Outside / Inside A 30°F (−1°C) 68°F (20°C) Brick / Air Space / Brick / Radiator	12 (38)	The normal loss through a typical brick cavity wall is shown in 'A'. Hot water radiators are often used for offices, schools, etc. and usually placed against outside walls. The inner face of the wall can rise to over 100°F (38°C) behind the radiator and 'B' shows the increased heat loss.
B Insulation	21 (66)	If insulation, as a board, or bright reflecting surface (preferably both) is placed behind the radiator, as in 'C', the loss is reduced. For a new building this insulation barely increases the cost if added during construction.
C	15 (47)	10ft² (approx. 1m²) so insulated will save over 1 gallon of oil per year (say 50p) and radiator sizes can be slightly reduced.

237

are inside the building being heated. However, if the mains and branch pipes to individual heating units are at a high level, as they usually are, there is some case for insulating them. The bare pipes radiate heat in all directions, and less than half of the radiation is effective at working level. Heat is also lost by convection, and all this loss goes into a rising stream of warm air. The major portion of the heat lost from the pipes is thus raising the temperature of the building structure and air, above working level. The temperature gradient from roof to floor is thus increased, and heat losses due to air change by roof ventilation and transmission through the material of roof roof, roof glazing and upper walls are also increased.

In some large buildings, especially of the single storey engineering type, the distribution piping can be giving out 10% of the total heat output of the heating system, and only a small proportion is actually improving the temperature at working level. Any reputable heating engineer positions his heating units to try to reduce the temperature gradient as much as possible, and it ought to follow logically that the distribution piping should be carefully insulated. It is appreciated this may mean a slight increase in the amount of heating equipment to offset the proportion of the heat loss from the piping which succeeds in doing some useful heating at working level. However, this extra capital cost will be recovered by steam saved every hour the system works.

With the fairly recent development of "radiant strip" type overhead heating, an alternative treatment of overhead distribution pipes is to convert them to radiant strips. The insulated reflector stops much of the upward warm air movement, deflects almost all the radiation towards the working level (Dia. 5A). Consequently the pipes can be made to give most of their heat loss as a contribution to the space heating system. Similar remarks apply to heating systems using high pressure hot water instead of steam, although with low pressure hot water piping it is possibly preferable to insulate, since radiation effect is fairly low compared to the convection loss. With steam systems the condensate return lines should be given similar consideration, i.e. insulated or made into radiant strip units.

238

CRITICAL APPRAISAL OF A DISTRIBUTION SYSTEM

We have considered many of the factors involved in the design, layout and operation of a heat distribution system using water or steam. How can these be applied to an existing system, and how can we most readily identify possible sources of saving? The following is a suggested sequence for a critical appraisal of a system.

SYSTEM LAYOUT. It is rare to find a plant engineer who has an up-to-date layout of the distribution system. With luck there may be a drawing of the original system, but this has almost certainly been outdated by modifications and additions. The starting point must be to bring this drawing up to date, or if no drawing exists, then to provide a new one. The drawing need not be elaborate — a single line drawing on an existing works plan will often suffice. However it should give a reasonably accurate layout, it should include valves, fittings and pipe sizes, and it should be capable of being used to record future alterations and modifications so that the same situation does not arise again.

INITIAL CHECK. The above exercise will almost invariably bring to light redundant piping and basic illogicalities of routing which can be immediately dealt with. It is not unusual to find a hundred feet or so of steam piping, previously used to feed some long abandoned machine or process, which remains live and terminates only in a steam trap. Heat is being wasted and unnecessary condensate being formed to overload the condensate system, or worse still, be passed to waste. In one such exercise recently undertaken the works engineer concerned was able to substantially replenish his stock of spare valves and pipe fittings by removing them from redundant piping.

STEAM FLOWS AND PRESSURE DROPS. An assessment of steam flows throughout the system should next be made. Again this need not be a complex exercise. Much information can be gained from a single main meter, by the simple expedient of turning off all other users except the one being checked. Figures can be checked against manufacturers estimates of consumption for process plants. Care should be taken to identify maxima and minima, and whether continuous or

239

intermittent loads are involved. Steam pressures should be noted at strategic points, and the pressure drops used to check flow information, and vice versa.

Another interesting exercise is to carefully make sure all machines are shut off by their stop valves, and then leave steam mains open for an hour or so, measuring the steam flow by meter or drop in boiler water level. Again results are often surprising.

PIPE SIZING. The previous exercises will again illustrate basic faults and may well shed light on existing distribution problems. The information should be used to provide a check of the adequacy or otherwise of pipe sizes throughout the system. In the worst cases a need for a section of larger bore mains may be illustrated. In many cases, even if not economic to install new pipe, the information will provide an invaluable guide for future additions and modifications which may be necessitated by alterations to the process plant.

INITIAL DISTRIBUTION SYSTEM AND HEADERS. Special attention should be given to the initial parts of the distribution system, particularly where trouble is being experienced with "carry-over" from the boilers. A check on the velocity of steam through the piping and fittings immediately on the boiler often illustrates what is at least a contributory factor towards carry-over. The method of calculation is illustrated earlier in this section, in the paragraph dealing with pipe sizing, so is not repeated here. A check, in the manner suggested, on the maximum pressure which can genuinely be expected at the exit from the boiler house complex will often give a salutory result.

INSULATION. A detailed check of insulation standards should be made in accordance with principles already discussed. Physical damage should be rectified and protection provided. Weatherproofing should be checked, and surface finishes refurbished as required.

TRAPPING AND AIR VENTING. The system should next be critically surveyed in regard to the adequacy of the draining steam trapping and air venting facilities provided. Air venting

240

is of particular importance. Because it is not seen its effect is often ignored, but it may well be making substantial reductions in the performance of the process plant. Traps should be examined not only in respect of location, but for sizing and suitability of application. The exercise should include an examination to ensure that condensate is being collected from all available sources and not passed to waste.

On the other hand there is no point in collecting condensate in a receiver, if due to bad design or pump selection the condensate pump cavitates and fails to handle the condensate such this this overflows to waste! Similarly it doesn't make sense to return condensate to the boiler feed tank, if faulty level control allows hot feed to overflow to waste. This fault is very common.

FLASH STEAM. Any visible escape of flash steam should be noted and investigated. If the escape is inevitable then the overall conditions should be examined with a view to installing a flash system on the lines indicated earlier in this section.

REDUCING VALVES. Process plants should be checked by systematic reduction in pressure to ensure that they are working at the lowest optimum steam pressure. It should then be ascertained that any reducing valves involved are correctly sized and adjusted. In some circumstances it may prove desirable to install two reducing valves of differing sizes in parallel to provide optimum control conditions.

Choice of Control Valves

In many liquid heating processes it is desirable to raise the contents o fthe tank or vessel to the required temperature as quickly as possible and then to hold this temperature for some considerable time. In view of the wide variation in steam flow required, difficulties may arise if a single valve is used for control purposes.

As an example, a 2 in. globe valve when fully open has a lift of roughly ½ in., giving an annular slot around the seat roughly equal to the seat area. If it is necessary to cut the steam flow by half, then the pressure drop through the whole pipework system to the vessel must be increased four times. This is done by closing the valve, and calculations for a typical

241

branch pipe from steam main to vessel show the final position of the valve will be only "cracked". Regulation is very difficult at such a small opening, and also as the steam is passing at very high velocity across the seat, scoring or "wiredrawing" occurs and damages the seat. This damage can cause steam leakage when the plant is not in use, and may take place to such a degree that the leakage is in excess of the "simmering" requirements of the process, so that frequent valve regrinding or replacement may become necessary.

One solution is to fit a much smaller valve in a by-pass round the larger one. The larger one can then be shut completely as soon as the desired temperature is reached, and "simmering" controlled entirely by the smaller valve, which can be kept fairly well open, eliminating any "wiredrawing" troubles. As Sir Oliver Lyle has said: "Many a reducing — or control — valve has been cursed and thrown out simply because it was too big."

Some automatic control valves have specially shaped ports to give accurate control over small flows while still allowing large flows when fully open. Other types operating with diaphragms and using air under pressure as the controlling medium often utilise two valves on parallel steam lines. Both are fully open until the desired temperature is nearly reached and the control air pressure starts to rise. The larger valve has such a weak opposing spring that it closes immediately, leaving the smaller one to modulate under the action of the controller. But having said all the above, control valve makers usually know their stuff! The trouble is that very often the client doesn't know the conditions sufficiently well. Take a look at the following figure which is typical of the flow conditions on a multi-cylinder drier; the valve has to cater for a wide range of conditions. It is of great value to be able to show the control valve makers the sort of conditions which his valve will have to cope with, and if this was done more often there would be less faulty controls!

HOT WATER SYSTEMS. While some of the checks suggested above apply only to steam, several of them apply equally to hot water systems. The main factor particularly applicable to hot water systems of all types is the maintenance of the correct balance throughout the system. In order to keep a regular

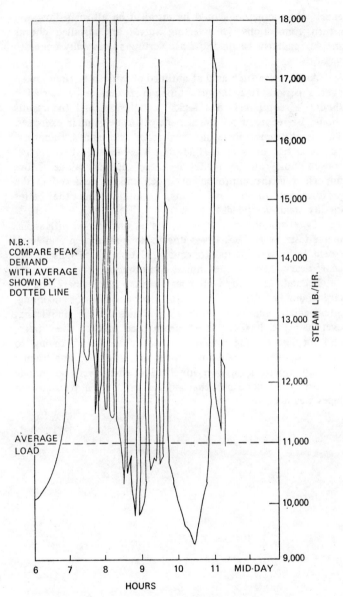

N.B.:
COMPARE PEAK
DEMAND
WITH AVERAGE
SHOWN BY
DOTTED LINE

STEAM LB./HR.

AVERAGE
LOAD

6   7   8   9   10   11   MID-DAY

HOURS

SHOWING ACTUAL METERED DEMAND ON AUTOCLAVES
Dia. 6

243

check, thermometers should be installed on all main flow and return connections. If diverting valves are installed checks should regularly be made with diverting valves fully open.

An exercise such as that outlined above can be undertaken over a period. It is important however, that, once started it should be continued to a logical conclusion, and the results should be recorded for the benefit of future similar exercises. The results must obviously vary from one plant to another, but it is not unusual to find that a saving of 10% of fuel consumption can be made by these methods alone. More important is the improvement which can be achieved in the performance of process plants, resulting from the better quality steam supplied.

Two examples can be given; first at one works, with steam meter tests by NIFES, losses due to poor insulation and small steam leaks were found to equal 215 tons of coal usage per year; nearly 13% of the annual consumption.

Second, at a large chemical works, where there were a large number of storage tanks all over the site for incoming, intermediate and final production liquids it was found that over half the 2000 kg/hr of steam used to heat these tanks and for line tracing was passing to unused steam mains, to heating coils in old tanks no longer in use, and to line tracing of dead pipework or carrying liquids which were so mobile at normal temperatures that no heating of the transmission pipes was needed.

9
Savings in Process

Methods of reducing process heat demands are discussed and simplified examples based on actual case histories are included to illustrate the principles involved.

DIRECT FIRED PROCESSES

The importance of reducing the amount of excess air in combustion has been stressed in previous sections of the Handbook under Burners and Furnaces. This is equally important in fuel fired processes, and by careful attention to optimising CO_2 values of the flue gases the heat demand can be significantly reduced. On larger plants, consideration should be given to the use of modulating burners instead of the on-off type especially on application where there is wide variation in heat demand.

In certain drying applications using oil or gas fired furnaces large quantities of excess air are introduced at the burner to provide low temperature flue gas which is directly circulated through the Drier. This practice can often lead to combustion problems such as flame chilling, smoke, etc., and is wasteful in fan power. Any dilution air should be mixed after the combustion zone. Diagrams 1 and 2 show the variation of flue gas losses for different values of CO_2 and temperature.

Example

A stentering machine (with heat exchanger) in a textile factory was found to have an oil consumption of 90 kg/hr. The setting temperature was 200°C and the exhaust gas temperature was 500°C. CO_2 concentrations of the flue gases averaged 6%. After the burner had been modified, it was found that the air fuel ratio could be adjusted to give a CO_2 of 11%. Simple baffles were introduced into the heat exchanger and the exhaust temperature dropped to 400°C. The oil consumption was reduced to 82.5 kg/hr.

247

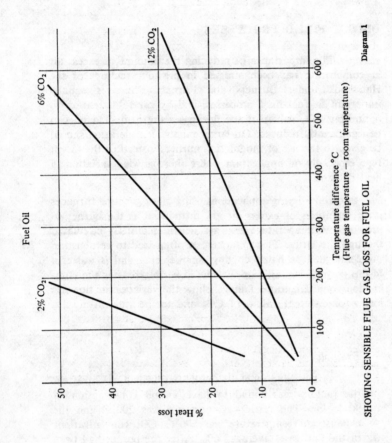

Fuel Oil

12% CO₂

6% CO₂

2% CO₂

% Heat loss

Temperature difference °C
(Flue gas temperature — room temperature)

SHOWING SENSIBLE FLUE GAS LOSS FOR FUEL OIL

Diagram 1

OPTIMUM LOADING AND SPECIFIC CONSUMPTION

Since the standing losses on many processes are more or less constant for a wide variation of loading rates, it will be realised that maximum economy will be achieved at the highest practicable throughput rate. It follows that the minimum number of units, working within their rated outputs, should be employed to meet production schedules. The optimum loading rate, which is not necessarily that given by the manufacturer, should be determined by comparing different throughputs with the equivalent fuel, steam or water consumption. The economic rating will be at the minimum cost per unit of product, i.e. the specific cost processed. Similarly, on batch feeding processes, the specific consumption and the process time gives an indication of the efficiency of operation. Where it is not practicable to maintain maximum loadings, the use of compartmented heating zones has proved to be economic.

Example

A sterilising machine operating at 115°C was found to consume 820 kg steam for 3400 litres of product which was equivalent to a specific steam consumption of 0.24 kg/litre. Experience of similar plant suggested that this consumption was high and investigations were carried out. The discrepancy was due to poor steam distribution and defective air vents, which when corrected, permitted the machine to operate at a lower steam consumption and the process time was reduced from 34 to 28 minutes, the specific consumption being 0.19 kg steam/litre.

Any methods which reduce the production time will also reduce the specific heat demand as well as the more obvious savings such as labour costs, etc. Such methods include the pre-heating of liquids and the product which will reduce the heating up period and thus the total production time. Of course, if this pre-heating can be achieved by waste heat recovery, from the process product of effluents, even more significant savings are possible.

Devices which increase the heat transfer to the product (such as agitators, stirrers, fans, etc.) can also improve production times. Reduction of liquid volumes and thus external surface areas of vessels will also lower the heat demand.

249

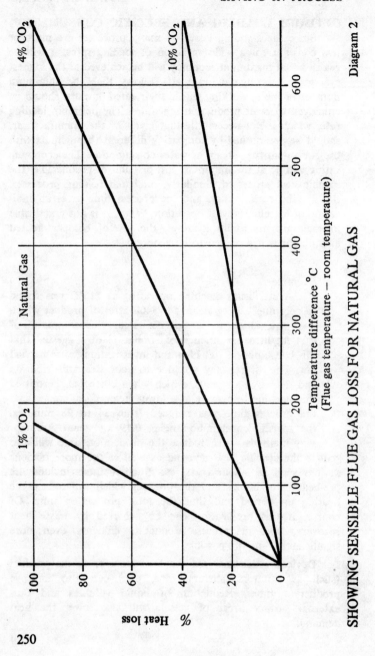

SHOWING SENSIBLE FLUE GAS LOSS FOR NATURAL GAS

Diagram 2

INTERMITTENT OPERATION

In many processes it is possible to reduce the heat demand if the heating medium is turned off during periods of non-production, batch feeding procedures, meal breaks, etc. It does not necessarily follow that attendant lengthy warming-up times will be involved.

Example

At one large laundry the steam flow during normal working to the finishing department was metered at 1400 kg/hr, and over the lunch hour this dropped to 500 kg/hr, with idle machines. By experiment it was found that if the main inlet valve was shut for 45 minutes, re-opening 15 minutes before re-starting the department, a steam saving equivalent to 10 tons of coal per year could be made. Diagram No. 3 shows the steam loading of the process.

As another example, at a textile works, cloth was dried and set to width in a stenter which essentially was a fan/steam heater battery system. It was shown that with the fan stopped the steam usage dropped from 1100 kg/hr to 150 kg/hr over the lunch hour, and that in practice there was no advantage in turning off the steam, as re-warming took about as much steam as the lunch-hour idling usage. However, during the metering exercise previously unsuspected peak steam demands of 1700 kg/hr were found during operating and these occurred due to interruptions of a few minutes in cloth flow, which appeared so short that the operative did not bother to stop the fan. When no cloth was passing over the air jets, the resistance to air flow dropped considerably and so more air passed through the heater battery to increase steam usage. A simple interlock was fitted to stop the fan when the cloth conveyor was stopped. This reduced steam usage by 1800 kg/week as stoppages averaged over three hours per week. Diagram No. 4 shows the steam loading pattern.

251

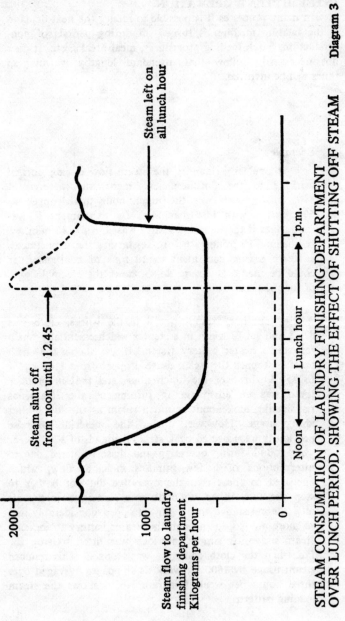

Diagram 3

STEAM CONSUMPTION OF LAUNDRY FINISHING DEPARTMENT
OVER LUNCH PERIOD. SHOWING THE EFFECT OF SHUTTING OFF STEAM

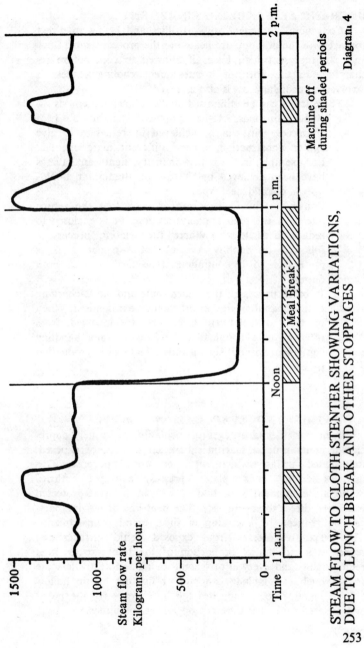

STEAM FLOW TO CLOTH STENTER SHOWING VARIATIONS,
DUE TO LUNCH BREAK AND OTHER STOPPAGES

Diagram 4

Machine off
during shaded periods

MECHANICAL HANDLING EQUIPMENT

Conveying equipment, tubs, trollies and bogies in direct contact with the product, are heated to the process temperature and can represent high losses if allowed to cool before recharging. Furthermore, if a subsequent cooling process is involved, a secondary loss is also incurred.

These losses can be minimised by the following methods: —

1. Reduce the mass of the equipment and thus the heat absorbed. This can be achieved by re-design to give lighter construction, use of different materials, etc. For example, in the bakery industry, significant savings have been achieved by the use of aluminium moulds instead of mild steel types.

2. Conveyors should be arranged such that the return line is in the high temperature zone or is enclosed to reduce heat losses. Where the heating process is followed by cooling, the use of two conveyors is preferable to a continuous type dealing with both duties.

3. In batch processes, the return route and the re-charging time of handling equipment should be minimised. When products of relatively large size are involved, it is often possible to unload and reload the same handling equipment without significantly affecting the production time.

RADIATION AND EVAPORATION LOSSES

Although the advantages of insulation are widely appreciated, many process machines have large areas of exposed, uninsulated surface which serve no useful purpose. This includes drying cans, platen presses, conveyor returns. Stationary equipment is easily dealt with by securing protected insulation direct to the surface. The heat loss of moving parts can be reduced by enclosing in sheet metal compartments. The evaporative losses from exposed liquid surfaces can represent a substantial proportion of the total process heat requirements and every effort should be made to provide some cover which can include canopies or floating plastic hollow balls. Diagram No. 5 indicates the heat losses from exposed and covered tanks for different operating conditions.

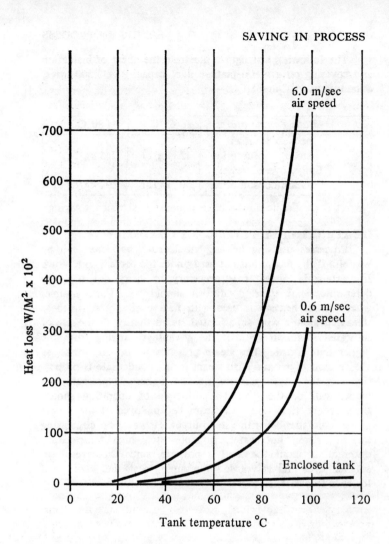

HEAT LOSS FOR EXPOSED SURFACES
FOR AMBIENT CONDITIONS
15.6°C & 70% SATURATION

Diagram 5

The following test figures illustrate the effect of insulation and covering on a metal plating plant capacity of 3,500 litres, operating at 65°C and 80°C: —

Bath Temperatures	65°C	80°C
Uncovered vessel steam consumption	123 kg/hr	265 kg/hr
Fully covered vessel steam consumption	91 kg/hr	191 kg/hr

HEATING MEDIA AT CORRECT CONDITIONS

Processes utilising heating media such as steam can be wasteful if the medium is not used under the correct conditions. In certain instances it is necessary to carry out trials to determine what these conditions should be. For example, where steam heating is used with heat exchangers, the heat release is slower with superheated steam than with the use of saturated steam in spite of the advantages arising from the higher temperature. The steam pressure is also important, as the amount of wasted flash steam in the condensate is proportional to the steam pressure, i.e. the higher the pressure, the higher will be the loss. The principle of operation should therefore be that saturated steam be employed at the lowest pressure compatible with rated output. However on circulating systems, using hot water, hot oil, etc., high temperature differentials should be used if possible, since these result in smaller pipes and valves, lower pumping costs and lower heat losses in transmission pipes.

Example

A direct steam injection dyeing machine consumed 900 kg/hr with steam at 7 bar pressure and the production time was 60 minutes. Reducing valves were installed and the steam pressure was dropped to 2 bar. It was found that the operation of the machine was considerably improved and the release of steam from the surface was minimised. The production period was reduced to 50 minutes and the steam consumption dropped to 820 kg/hr.

Example
Agitators in a chemical works were consuming 12 m³/min. of compressed air at 7 bar line pressure. It was found that the air was reduced to 0.7 bar for use on the machine. The air supply was transferred to a low pressure compressor with resultant power saving of 50% with no production difficulties.

PRE-DRYING EQUIPMENT

It will be appreciated that physical methods of liquid removal are much more efficient than evaporative processes. For example, when processing newsprint, it has been shown that moisture removal during the initial mechanical drying operation is eight times cheaper than the subsequent evaporative process. Mechanical driers include filters, mangles, and centrifugal hydro-extractors. Operating conditions for these machines should be set to ensure optimum overall drying efficiency and the performances checked by batch weighing. Any inefficiency in this pre-drying equipment will inevitably increase production time and heat demand in the drier proper.

FINAL MOISTURE CONTENT

The final acceptable moisture content should be determined as it is clearly uneconomic to dry below this value. This is normally the equilibrium moisture level, i.e. the moisture level which the products attain in storage or packaging with normal atmospheric conditions. However, in certain applications where subsequent operations are involved, such as pulverising, lower levels may be specified. To illustrate the effect on heat requirements, assume that a substance is dried from a moisture level of 25% to 0.5% (dry basis). The heat demand is 1400 kJ/kg material which includes drier radiation and convective losses. If it were possible to dry to the equilibrium moisture level of 5%, the heat requirement would be reduced to 1230 kJ/kg, i.e. a saving of 11%.

Overdrying is common and there can be no doubt that this is one of the greatest factors which affects thermal efficiency and continuous determination of final moisture should always be considered. There have been many developments in this direction including Infra-red absorption, Micro-wave

257

absorption, Conductance, etc., and the SIRA Institute of Great Britain have developed an automatic control technique based on measurement of the difference in temperature between the material and the wet bulb.

THE PROCESS OF DRYING

Substances to be dried usually have free moisture on the surface and when subjected to drying processes, exhibit three distinct stages in the rate of evaporation: —

1) Initially, the drying rate increases as the wet material is raised in temperature.

2) During the second stage, the rate remains substantially constant as water evaporated at the surface is replaced by water from within the substance. The temperature of the product remains constant. In convection driers, this temperature is virtually at the wet bulb temperature of the air. In conduction driers the solid gradually attains the boiling temperature of the liquid.

3) When the surface becomes relatively dry, the rate of evaporation progressively decreases, the stage being known as the falling rate period. The temperature of the product usually increases.

Most materials when presented to the drier show wide variations in the duration of each stage, certain products having virtually no "constant rate" characteristics.

Diagram No. 6 shows drying curves obtained from a brick drier.

For the purpose of this Section, two types of driers are considered, the convective type and the conduction type. In the convective (or direct) type, the hot gases are brought into direct contact with the product and carry away evaporated moisture. In the conduction (or indirect) drier, heat is transmitted to the product by conduction (and radiation) and moisture is removed independently of the heating medium.

The thermal efficiency of a drier is the ratio of the heat required to evaporate the moisture in the product to the total heat input to the drier (which includes radiation losses).

258

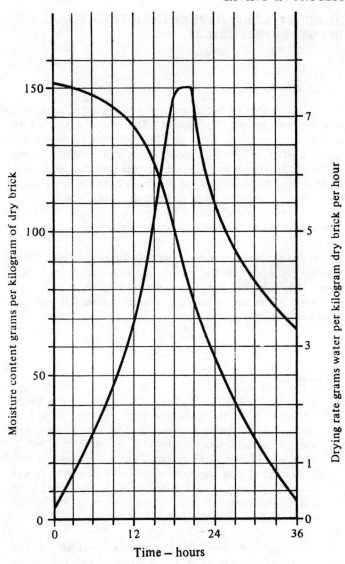

DRYING RATE AND MOISTURE CONTENT
FOR TYPICAL BRICK DRYER

Diagram 6

HUMIDITY, TEMPERATURES AND VELOCITIES IN CONVECTION DRIERS

a) *Humidity*

For maximum efficiency the drying air should be exhausted at the maximum possible humidity. During the constant rate period, evaporation is rapid and the full flow of drying air can be discharged with high moisture burdens. However, as the "falling rate" condition is reached, high exhaust losses will occur if humidity control is not practiced. Furthermore, certain products such as pastes or pulps form dry crusts which impede moisture removal and can result in cracking and shrinkage problems if external humidities are kept too low.

These problems can be overcome by recirculation of a variable proportion of the exhaust air to the air inlet to the drier. This principle not only limits the exhaust losses but also offers a measure of humidity control, which can result in more uniform drying.

Example

It was found that the exit air conditions in a single pass convection steam heated drier were 43°C with a relative humidity of approximately 50%. The exhaust flow was 620 m³/min with an air inlet temperature of 93°C. 50% recirculation of the exhaust gas was introduced and it was found that the temperature and humidity of the exhaust increased to 46°C and 75% R.H. The drying time was unaffected and the steam requirement of the heater battery was reduced by 10%.

Indicating and recording hygrometers installed on the exhaust system will lead to the maintenance of high efficiencies and will also indicate variations in operating conditions for different materials. Table 1 indicates the effect on steam consumption by increase in exhaust humidity.

260

TABLE NO. 1 — TESTS ON BATCH DRIER

*Showing the effect on Exhaust Relative Humidity
and Steam Consumption*

Time from Start of Cycle	Test No. 1		Test No. 2	
	Inlet Air Damper Position	% R.H. of Exhaust	Inlet Air Damper Position	% R.H. of Exhaust
0-1 hour	shut	75	shut	80
1-6 hours	¼ open	70	½ open	64
6-12 hours	½ open	20	⅜ open	45
12-18 hours	¾ open	15	¼ open	32
18-24 hours	full open	0	¼ open	15
Steam used per kg. moisture evaporated —kg	4.28		2.37	

b) *Gas Velocities*

The rate of evaporation is dependent on the thickness of the air or gas film surrounding the product. This film thickness is reduced with high air velocities and the use of fans, slots and jets discharging high speed air will promote higher drying rates. Diagram No. 7 indicates the drying curves of a typical tray drier and shows the effect of different air velocities on drying rates

If velocities vary in different parts of the same drier, for example if Dia. 7 shows air velocities over different trays, then in order to obtain an acceptable moisture content of material on the tray over which the lower air velocity exists, over-drying must be occurring on the other trays. By equalising air velocities, using baffles, drying time can be reduced, increasing useful throughput and reducing steam usage per kg. of dried product.

If high air velocities exist, checks should be made to ensure that the air makes good contact with the product to ensure optimum moisture pick-up.

Diagram 7

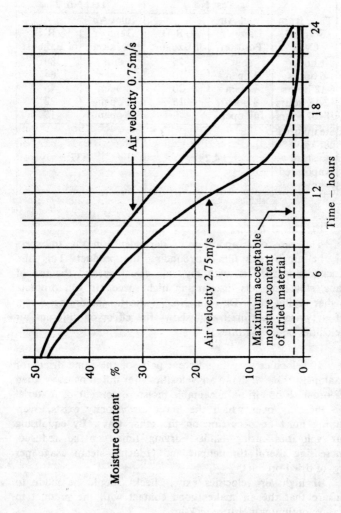

AVERAGE MOISTURE CONTENT OF MATERIAL
DURING A CHANGING CYCLE IN A TRAY DRIER

Evaporation rate is related to the exposed area of the product being processed. Methods to increase exposed areas include granulating and extending of materials, use of thin sheets and reducing the blank supporting area under the product during processing.

c) *Temperatures*

The operating temperature should be as high as possible consistent with good product quality. For continuous dryers, the principle of counterflow should be practiced, i.e. the product travels in the opposite direction to the drying gases. This arrangement ensures higher temperature differentials leading to higher drying rates. Many firms are worried about using high air temperatures at the wet end, due to fears of damage to the product, but in fact, while above the critical moisture level, the material can only reach the wet bulb temperature of the gases. For example, a continuous drier had an exhaust air temperature of 66°C with 30% R.H. However, if this exhaust temperature were increased to 120°C the wet material passing in counter current flow would not exceed a temperature of 88°C if the same relative humidity of 30% were maintained. However, the resulting heat content in the exhaust air would be 30% lower and there would be no loss of drying rate.

CONDUCTION (OR INDIRECT) DRIERS

a) *Heat Transfer Rate*

It will be appreciated that increases in heat transfer will result in faster drying rates, lower specific energy costs and lower capital costs in the drier installation. The maximum heat transfer will be restricted by the nature of the material being processed, the temperature of heating medium available, degree of agitation, the film thickness of the heating medium and the thickness and conductivities of the heat transfer surface.

b) *Heating Mediums*

Reference has already been made for the need to use heating mediums at suitable conditions for the process, e.g. Steam for cylinders, platens, etc., should be supplied at dry saturated conditions. The use of superheated steam is associated with lower heat transfer rates and erratic drying.

263

Example

A cylinder drier operated at 10 bar with 30°C of superheat consumed approximately 9,000 kg steam/hr. Following the installation of new boiler plant, saturated steam at the same pressure was supplied to the plant. It was found that a 10% increase in machine speed was possible for no increase in heat demand.

To maintain high transfer rates and to reduce power consumption, it is essential to ensure that the condensate is removed as quickly as possible. On slow speed (below 300 m/min) it is usual to use stationary siphon pipes which dip into the pool of condensate at the bottom of the cylinders and regular attention to the pipe clearance and steam traps will result in minimum retention of condensate in the cylinder bottoms.

In trials on a paper machine it was shown that the electrical power load was increased by 20% due to water-logging.

On high speed cylinder driers (above 300 m/min) it has been found that rotating syphon pipes have maintained a condensate layer of less than 1 mm thick, resulting in higher heat transfer rates and better stabilisation in the cylinder drive.

It is often advantageous in multi-cylinder and multi-pressure driers to cascade the condensate from the high pressure cylinders to flash vessels, which thus supply lower pressure steam to the remaining cylinders. Flash steam losses in condensate lines are considerably reduced.

Attention to water treatment and venting will ensure minimum quantities of air and gases in the steam and reductions in scale and rust deposits in the cylinders, both of which can change drying rates considerably.

Bad trapping, fouling of surfaces and poor steam entry arrangements may cause variations in heat transfer across the surface of drying cylinders and so cause variations in temperature of material such as cloth or paper being dried. Diagram No. 8 shows measured temperatures of material passing over the multiple cylinders of a drier. It will be seen that due to bad steam circulation, poor air-venting and poor condensate removal, temperatures of material at the front (steam admission side) of the cylinders was higher on every cylinder than at the back (condensate removal side). Successive modifications enabled

Diagram 8

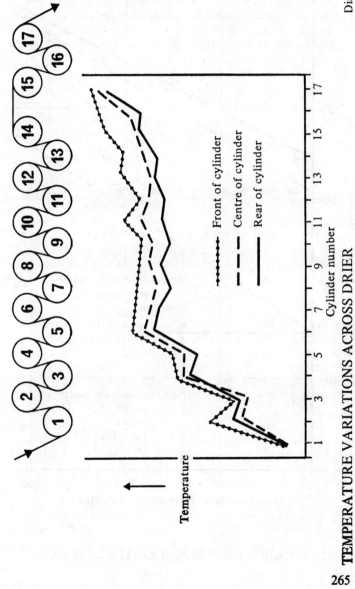

Front of cylinder
Centre of cylinder
Rear of cylinder

Temperature

Cylinder number

TEMPERATURE VARIATIONS ACROSS DRIER

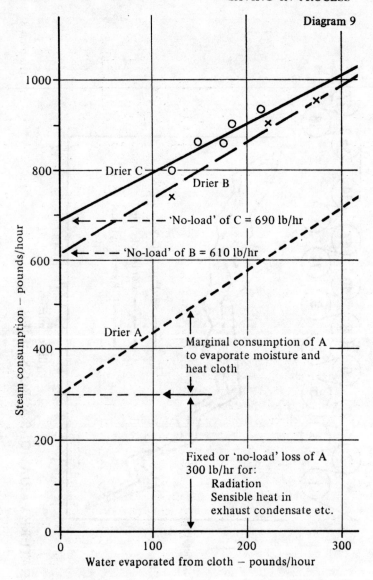

Diagram 9

PERFORMANCE OF THREE TEXTILE DRIERS

these differences to be eliminated and the drier could then be operated at 10% higher throughput for the same steam consumption. Originally steam was wasted by over-drying material at the front ends of the cylinders in order to obtain satisfactory drying at the rear.

Driers, like many other items of process plant, usually have high "no-load" energy usages. Diagram No. 9 shows test results obtained on three cloth driers, and it will be seen that the "no-load" steam consumptions range from 45 to 70% of full load usage. It is therefore important to keep such driers fully loaded, even if this means organising production so that drying only takes place for possibly half the shift time and the drier is then shut down when not needed.

For example, drier 'A' was found to take 1000 lbs. of steam to heat up to normal working temperature. If run at full load for four hours it would remove 1200 lbs moisture from cloth using 2800 lbs. steam. However if run at half load for eight hours, removal of the same weight of moisture would use 4000 lbs. The respective steam usages, including heating-up are thus 3,800 and 5000 lbs, or a 32% increase in steam consumption.

c) *Agitation*

The heat transfer to the product depends on the degree of agitation of the product relative to the heating medium. Agitation reduces film thicknesses and causes rapid mixing of hot and unheated products.

This effect is achieved by cascading, rolling, and in the case of a direct fired dryer, turbulence in the combustion gases. On the product side methods used to increase agitation include pumps, stirrers, fans, vanes and baffles.

Example

A pan drier processing fertiliser by dehydration had a production time of 24 hours and consumed approximately 500 litres fuel oil. It was found that the heat transfer was impeded by the product caking on the heating surfaces and poor turbulence and distribution of the flue gases. The flue gases were recirculated and baffles were introduced in the flue ways. The production time was reduced to 11 hours and the fuel consumption dropped to 380 litres.

d) *Hooding of Cylinders*

The provision of vapour hoods over drying plant are used to confine and control the humidity of the air leaving the machine. The rate of extraction, if controlled by humidity measurements, will limit the exhaust losses.

Many large paper drying machines are now equipped with enclosures which encase the whole of the machine. The supply and exhaust air are balanced within the hood itself and only a small proportion of air is taken from the shop. Input air heat exchangers can be incorporated to extract heat from the exhaust.

The advantages of closed hoods include reduced steam consumption and lower fan power requirements. Tests on similar machines indicated consumption of 3000 kg. steam per tonne of paper processed for an open-hooded arrangement, compared with 2,500 kg steam per tonne for a closed hood with heat recovery system.

WASTE HEAT RECOVERY

Waste heat recovered from vapours, flue gases and effluents can be utilised in two forms: —

1) by direct recovery and return to the process
2) by indirect heating of a secondary heating medium for use on a different process.

Direct recovery has the advantage that the heat recovery is in phase with the process. Consequently, it is often possible to design systems such that the process can always accept the recovered heat and difficulties resulting from intermittent operation are eliminated. However, this method suffers from the disadvantage that the amount of heat recoverable is limited by the process requirements.

For example, where preheating of combustion air is practiced, the amount of heat recoverable is often less than 50% of that theoretically available.

The amount of heat recovered in an indirect system is a function of the heat requirements of the secondary process. By careful selection of a secondary process with a low grade heat demand, very high overall efficiencies can result. Problems arise when the main process has a variable or intermittent loading pattern and in many cases it is necessary to provide heat storage arrangements to offset the variations in heat recovery.

268

Diagram 10

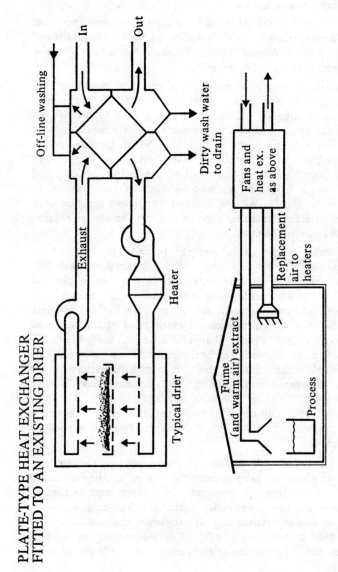

PLATE-TYPE HEAT EXCHANGER
FITTED TO AN EXISTING DRIER

In

Out

Off-line washing

Dirty wash water
to drain

Exhaust

Heater

Typical drier

Fume
(and warm air) extract

Process

Fans and
heat ex.
as above

Replacement
air to
heaters

HEAT EXCHANGER/PROCESS VENTILATION

269

The major requisite for successful heat recovery is the heat exchanger which must meet the conditions imposed by the fluids being handled in terms of chemical composition, temperature and flow rates.

Fouling of the exchanger surfaces is a very important consideration and the system design calculations should incorporate suitable fouling factors. Provision for routine cleaning and maintenance is also important.

Diagram No. 10 shows the application of plate-type heat exchangers to two process systems. The first is a drier, where the warm outgoing exhaust air is being used to preheat the incoming air. The second is an extract system removing vapour and fumes from process vats. While this ventilation may be essential to give acceptable working conditions, it does take large quantities of warm air out of the factory building and this volume has to be replaced by fresh air which has to be heated to give reasonable comfort conditions for the work force. The heat exchanger shown can cut the energy requirements by more than half.

While tubular exchangers can be used, plate-types have advantages as they can more easily be cleaned by wash jets. Such cleaning is essential to remove dust, grease or material particles which would otherwise build up to reduce heat transfer and increase resistance to air or gas flows. The plates can be of metal, aluminium for clean gases or stainless steel for more corrosive gases. Glass plates are also used. Plates can recover some of the latent heat from moisture-laden gases by allowing condensation on them, but precautions have to be taken to ensure good drainage and prevent thick films of moisture building-up and hindering heat transfer. The heat "wheel" operating on a regenerative cycle, is well-known in air conditioning work, but is being increasingly applied to process use.

The recoverable amounts of heat depend on the temperature to which the exhaust from the process can be reduced at the heat exchanger. A temperature difference must be allowed between the two fluids in the exchanger. The economic limit (i.e. to give the economic heat exchanger size consistent with low fluid transmission costs) will vary according to the processes being considered. For example, on dyehouse effluents

270

10° to 25°C is often used. Whereas in diesel exhaust systems, temperature differentials of 40° to 50°C are quite common.

WASTE HEAT RECOVERY FROM FLUE GASES

Typical exhaust temperatures of a number of different processes are listed in the table below, together with average heat content of the flue gases.

	Exhaust Temp.	*Exhaust Heat Loss*
Diesel Engine	400°C	35%
Glass Tank	500°C	30%
Lancashire Boiler		
(without economiser)	450°C	28%
Textile Stenting Machine (indirect)	450°C	28%
Food Oven (indirect fired)	500°C	35%
Modern High Efficiency Boiler	230°C	12%

Since most fuels produce exhaust gases containing sulphur dioxide and water vapour, corrosion of heat exchangers can occur unless this factor is given serious consideration. Modern materials such as stainless steel offer the solution to many corrosion problems but this can be expensive and is not justified for certain applications. At the expense of some recovery, cheaper materials can be used in such cases, provided the design is such that corrosive conditions do not develop.

This can be achieved by ensuring that metal temperatures in the exchanger are maintained above the acid dew point i.e. approx. 145°C for fuels up to 4% sulphur.

Example of Direct Recovery:

Furnaces processing an expanded insulation had an exhaust temperature of 760°C with an oil input of approximately 150 litres/hr for each furnace. The flue gases were used to convey the product from the furnace to the collection silos. It was decided to proceed with a heat recovery scheme but due to the intermittent operation of the plant, combustion air preheating was adopted, particularly as there was little scope for the utilisation of heat elsewhere in the factory. Due to the very abrasive nature of the product, conventional designs of heat exchanger (with high turbulence and velocities) could not be used. The final design consisted of a 20 ft length of 250 mm

271

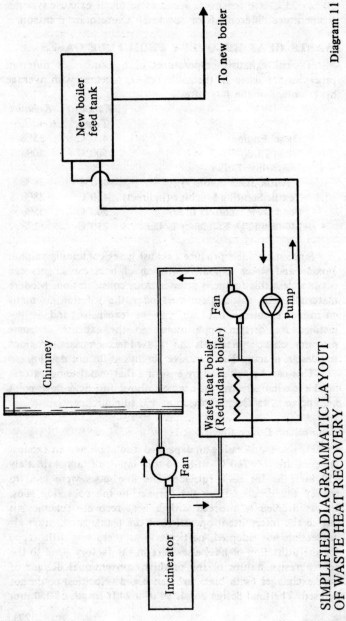

New boiler feed tank

To new boiler

Diagram 11

Chimney

Fan

Waste heat boiler (Redundant boiler)

Pump

Fan

Incinerator

SIMPLIFIED DIAGRAMMATIC LAYOUT
OF WASTE HEAT RECOVERY

finned steel tube over which passed combustion air en route to the burner. Due to the very low velocities, little abrasion took place. The combustion air was preheated to 180°C and fuel consumption was reduced by about 10%. Capital costs were recovered in less than one year.

Example of Indirect Recovery:

An industrial incinerator burnt 45 tonnes of waste paper per week and the average exhaust temperature was 425°C. It was decided to recover heat from the exhaust to provide feed water heating for the new boiler plant. Diagram No. 11 shows diagrammatically the arrangement. One of the old redundant boiler shells was modified and used as a low pressure heat exchanger. The feed water temperatures to the boiler were raised by approximately 25°C when in operation. The fuel consumption of the boiler plant was reduced by 2%, capital costs being recovered in 18 months.

HEAT RECOVERY FROM EFFLUENTS

Most effluents contain solids and corrosive chemicals. The selection of the materials for the heat exchanger surfaces must not only take into account the corrosive effects but also of the fouling characteristics of the materials.

For instance, some metals, whilst exhibiting full anticorrosive characteristics, can sometimes cause fouling problems due to deposits adhering to the protective film formed on the metal surface. Materials in common use include stainless steel, graphite and "teflon". Systems dealing with effluents containing high concentrations of suspended solids should incorporate suitable strainers, settling tanks, etc., to increase the availability of the plant between cleaning operations.

Since the available heat from effluents is usually low grade, the liquids involved require long flow passes to recover the heat efficiently. This is achieved by multi-passing in tubular and plate exchangers and by separation into smaller parallel flows in the case of graphite exchangers.

It is often considered that where effluents are relatively cool, say 30° to 40°C, heat recovery is not viable. However, it is sometimes possible to find a low grade heat demand such as incoming cold water, which can often be raised by 5° to 10°C. In certain cases, diversion of hotter effluents into separate

273

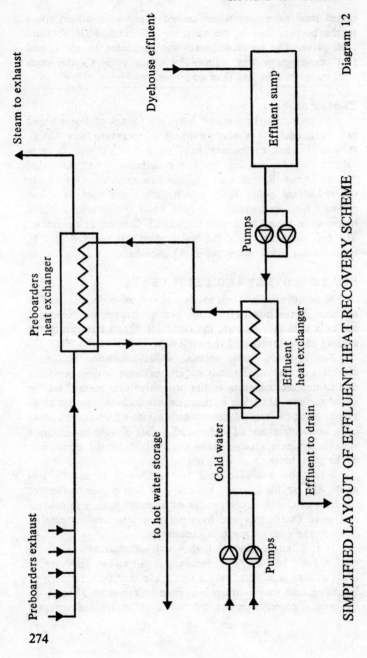

Diagram 12

SIMPLIFIED LAYOUT OF EFFLUENT HEAT RECOVERY SCHEME

tanks is practicable, and can show significant savings over systems recovering heat from mixed cooler effluents.

Example

Tests in a hosiery finishing factory indicated that the average dyehouse effluent flow was 300 litres/min at an average temperature of 140°F. Exhaust of steam from the pre-boarding department averaged about 700 kg/hr at a pressure of 0.137 bar g. Hot water requirements from the dyehouse were approximately 130 litres/min. A heat recovery system was installed consisting of two heat exchangers working in series to provide process hot water. The effluent heat exchanger was the stainless plate type, but, due to pressure loss considerations, it was found necessary to use a single pass tubular exchanger for the pre-boarding application. The actual performance of the system was such that the steam consumption was reduced by an average of 900 kg/hr and the boiler plant, which was previously overloaded at peak production periods, was able to cope with the maximum steam demand. Capital costs were recovered in less than 2 years. Diagram No. 12 indicates a simplified layout of the arrangement.

Diagram No. 13 shows an actual installation at a textile factory and illustrates several points already discussed. Warm effluent is passed through a stainless steel perforated sheet filter with continuously rotating brush to prevent fouling, and the sump formed in the floor serves as storage of the low-grade heat. When the level trips the float switch the pump passes the effluent through the heat exchanger and then to drain and cold water is preheated. This cold water is only passed when the pump is running and the throughput controlled to give a fairly constant discharge temperature around 30°C. This process water is further heated by passage through a coil in the process condensate collection tank. This method is adopted to ensure the condensate is cooled below 90°C to prevent "flash" steam losses and also enable it to be pumped without fear of cavitation or steam-locking in the pump. Condensate returned to this tank is passed to a perforated pipe in the bottom so that any "flash" is re-condensed. To reduce the amount of "flash" tending to be formed at this perforated pipe, the high-

275

Diagram 13

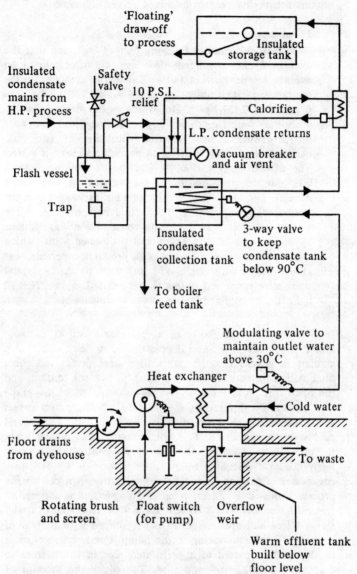

AN 'INTEGRATED' HEAT RECOVERY SYSTEM

276

pressure condensate from drying and crease-resisting processes is passed through a separate flash tank and the resultant low-pressure steam used to further preheat the process water.

This system eliminated clouds of steam lost from the previous collecting tank and enabled dye vessels and scouring processes to be fed with water at 40-45°C instead of mains temperature averaging around 10°C. The energy consumption of the department was reduced by 20% and the cost of the whole installation was recovered in about two years (at 1973 fuel prices!).

WASTE HEAT RECOVERY FROM VAPOURS

Heat recovery from exhaust vapours can show substantial savings since these vapours carry away a large proportion of energy absorbed in the form of latent heat. Condensing of these vapours by heat recovery not only shows financial return in fuel costs, but can also result in the reduction of pollution, noise, etc., and sometimes the re-use of the condensate thus produced.

In certain processes, where the condensate from the vapour is recovered, it is often difficult to find a use for this relatively low grade heat. In such cases, it is sometimes possible to use two stage cooling and condensing which permits a proportion of the heat to be recovered at a much higher temperature than the required final condensate condition. By selection of heat exchanger sizes, fuel savings can be achieved as well as reduced pumping costs of the coolant.

Diagram 14 shows two methods of possible recovery of heat from vapours.

Example

Steam exhaust flow from a steel forging shop amounted to approximately 9,000 kg/hr under working conditions and 4,800 kg/hr during hammer idle periods. Complaints had been received from the local health department about the noise of the exhausts and the amount of vapour fog produced particularly during humid days. The system of cylinder lubrication was changed such that heat exchangers could be installed to use exhaust steam without attendant fouling problems. The heat thus recovered was used for boiler feed water (which is usually

277

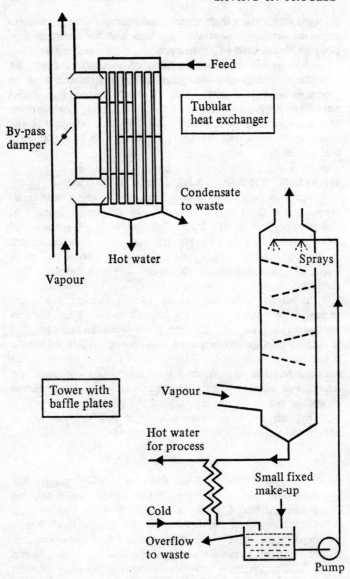

By-pass damper

Feed

Tubular heat exchanger

Condensate to waste

Hot water

Vapour

Sprays

Tower with baffle plates

Vapour

Hot water for process

Small fixed make-up

Cold

Overflow to waste

Pump

TWO METHODS FOR RECOVERY
OF HEAT FROM VAPOURS
(Both are successfully in use)

Diagram 14

278

at a very low temperature in forges) and space heating of adjacent buildings. Noise and the fogging problems were reduced. The condensate return not only reduced the heat demand of the boilerhouse, but also cut water and treatment costs considerably. Capital costs were recovered in two years.

Example

A solvent recovery plant condensed petroleum spirit vapour from 104°C to 15°C using cooling water at 11°C inlet and 30°C outlet temperatures. The flow of coolant was approximately 450 litres/min of which approximately one-third was used for boiler feedwater purposes and the remainder was circulated to cooling towers. An additional heat exchanger was installed in the vapour circuit prior to the original condenser and coolant was circulated through this at approximately 15.0 litres/min, inlet and outlet temperatures being 11°C and 70°C. The whole of this was used for boiler feedwater purposes and resulted in a fuel saving of approximately 5%.

Diagram 15 shows a summary of all the items discussed in Sections 8 and 9 as a reminder to the reader who may care to itemise the heat losses in his own factory or building in the heat distribution or process plant system.

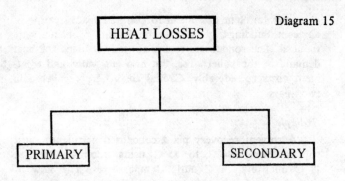

Diagram 15

HEAT LOSSES

PRIMARY

1 RADIATION
(pipes, furnaces,
buildings, hot surfaces)

2 LEAKS
(steam, water,
compressed air, etc.)

3 SAFETY VALVES
*blow-off losses
from boilers,
process plant, etc.*

4 FAULTY TRAPS
*wrong types
over-sizing
lack of maintenance, etc*

5 EXCESSIVE
VENTILATION

6 OVERHEATING
*absence of control,
faulty control*

SECONDARY

1 EXHAUST
(driers, pumps)

2 FLASH STEAM
(lost from condensate)

3 CONDENSATE
*discharging from waste,
overflowing to waste
from collection receiver*

4 BLOW-DOWN
*from boilers,
from process vessels*

5 VAPOUR
(pans, vats,
evaporators,
processes)

6 HOT EFFLUENTS
*waste liquors,
dye vats, etc.*

DIAGRAM 16

TOTAL WATER USAGE PROCESS FACTORY

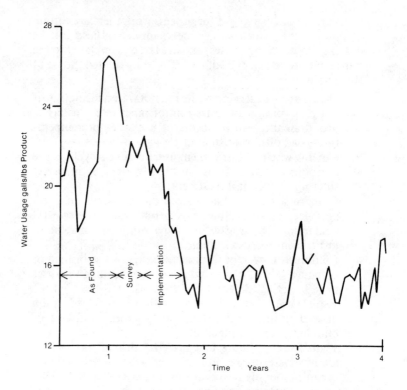

Water for use in process, etc. is costing more each year, and nationally we are approaching shortages in dry period — necessitating high capital expenditure to increase supplies which in turn must increase water costs. Detailed surveys can produce dividends, both by reducing cost of purchased water and also possibly by reducing costs of effluent treatment.

The survey should examine (a) excessive use for cooling (very small rise in temperature between inlet and outlet), (b) excessive use for washing and rinsing (by experiment determine the minimum necessary to remove chemicals, dirt or impurities), (c) wastage by operatives (taps, etc. left open), (d) re-use of water leaving one process as water for other processes, etc., (e) installation of cooling towers to reject heat and re-cycle water.

This diagram shows a large process plant using water for multiple purposes. The left hand side shows initial consumption cycle up to the start of the implementation of recommendations after over a year of metering and survey work. Consumption was reduced very quickly from an average of around 22 gallons/lbs finished product to 16 gallons. The metering is continued in later years to check for drift due to relaxations in the new operating methods.

SUMMARY OF SECTION 9, FOR USE ON ANY INVESTIGATION OF ENERGY USAGE IN PROCESS

At every firm using heat for process work there are certain common items which should be examined. These can be causing major energy losses, and are largely unrelated to the particular process or type of material being treated. Some of these points are:-

1. **Distribution of services** — steam, hot water, insulation of hot pipework, incorrect sizing of pipes, etc., leakages, bad drainage, use of incorrect pressures or temperatures, bad or non-existent metering.

2. **Cooling water** — extravagent use, possible recycling, use of cooling towers, re-use in other processes, incorrect filtering or chemical treatment.

3. **Compressed air** — badly sized compressors, incorrect pressure, leakages, bad water removal, excessive oil contamination, waste in idling tools or machines, insufficient receiver volumes.

4. **Combined power/process possibilities** — use instead of reducing valves, opportunities when extending factories or re-boilering. Steam drives for fans, etc.

5. **Plant fabric** — insufficient insulation, casual infiltration of cold air into driers, etc., also structural insulation of buildings, furnace plant, etc.

6. **Plant running hours and material throughput** — bad loading or excessive idling times waste energy. Better organisation of conveyors or material handling may improve plant loading. Incentives to encourage throughput or efficient energy use.

7. **Waste heat recovery** — from cooling air or water, from drier exhausts, from vapours, from process liquids or solids, from hot gases leaving furnaces, etc.

8. **Recovery from hot effluents** — recycle heat back to process, or use hot effluents as possible feeds to non-critical processes or heating elsewhere in factory.

9. **Condensate return** — from distribution, process and heating, flash steam recovery, good trapping practice.

10. **Electricity** — bad motor sizing; wasteful drives; incorrect and wasteful lighting. Spot heating usage, bad power factor, bad load patterns.

11. **Economies of alternative fuels for direct-heated processes; storage and distribution problems** — examination of combustion equipment and combustion conditions, and control of furnace atmospheres.

12. **Use of automatic controls** — advice on correct sampling of temperatures, gases, etc.
13. **Safety and noise problems, general environmental comfort of operatives.**

10
Small Industrial Furnaces

INTRODUCTION

Small furnaces are playing an increasingly important part in modern industry. Almost every manufactured article requires heat at some stage of production and well-designed and efficiently-operated furnaces can have a significant effect on cost, output and quality.

The term industrial furnaces covers a wide range of heating appliances where temperatures are raised for thermal processes including: —

Heat treatment of metals up to nearly 1100°C to produce specified physical or metallurgical properties.

Reheating operations up to about 1350°C for such processes as rolling, extrusion, forging and pressing.

Melting and refining of metals at temperatures of up to nearly 1700°C.

There are few small furnaces fired by solid fuel in existence but coke is used in metallurgical processes, for example, cupolas, both as a fuel and a reactant.

There are, of course, many electrically heated furnaces, but due to higher energy costs, electricity is only normally used for sophisticated furnaces where critical metallurgical requirements have to be met or when metal loss in melting operations is of vital significance. In these instances overall operating costs can favour electricity.

However, there are no combustion or waste gas losses with electric furnaces and providing heat losses are kept at a minimum and they are used efficiently there is little scope for reducing consumption. There is, however, no further specific reference to electric furnaces in this section.

In marked contrast to boiler plant, when efficiencies of over 80% are not uncommon, thermal efficiencies of industrial fuel fired furnaces generally range between 10% and 40% depending on temperature requirements, with only continuous furnaces reaching the higher figure.

Detailed investigations of furnaces have shown that many are operating well below optimum overall efficiency and substantial fuel savings can often be achieved by the simple application of basic principles.

In addition, when considering overall cost-effectiveness, improved fuel utilisation invariably results in increased productivity, lower rejection rates, lower maintenance costs and a better working environment. These latter benefits often far outweigh fuel savings.

There are many different types of small furnaces in use and the design and construction vary widely as do the operating conditions and duty they have to perform. Generally, the guiding principles of practice and operation are common to most of them.

Many factors affect efficiency and performance of furnaces and some of the more important ones are discussed under appropriate headings in the following pages.

CHOICE OF FUEL

The choice of the right fuel is of the greatest importance for every industrial process as it can have a vital effect on: —

Production costs
Work quality
Maintenance costs
Productivity
Working environment

It can be seen, therefore, that in considering fuel there are many factors affecting cost which must enter into the equation; the industrialist is concerned with overall costs as well as conservation. The basis of Fuel Management is measurement; measurement is necessary because it involves the establishment of targets, monitoring of performance, fuel conditioning, etc. It demands also a willingness to alter traditional methods to take advantage of potential fuel savings and in many cases will necessitate the setting up of special fuel sections.

Selection of Fuel

This immediately highlights the need to take cognisance of such items as: —

(a) Fuel or energy availability and future trends.
(b) Quantities required and negotiation of contracts with suppliers.
(c) Security of supplies and possibility of dual fuel or dual energy system and standby.
(d) Impact on preferential tariffs, e.g. group buying.

(e) Selection of plant.

(f) Viability studies or investment appraisal.

The vast majority of small furnaces are fired by gaseous or liquid fuels.

Gaseous Fuels

Included in this category are manufactured gas, liquid petroleum gases, propane and butane and natural gas.

Gaseous fuels are in the ideal state for intimate mixing with the minimum of excess air, gas and air can be accurately controlled to give the rate of heat liberation required with the desired furnace atmosphere.

Gaseous fuels are perhaps the most suitable of all fuels for application to heat treatment furnaces by virtue of their consistency of quality and ease of control. Their high purity renders them particularly suitable for special applications where sulphur would be injurious. The use of these fuels often results in reduced maintenance costs, extended refractory life and increased furnace availability.

Liquid Fuels

Liquid fuels used for furnace heating are gas oil, light and heavy fuel oils and coal tar oils. These fuels have many of the advantages associated with gaseous fuels.

Smokeless combustion of liquid fuels cannot be achieved without an excess of air and this makes it difficult to obtain the reducing atmospheres sometimes required in practice.

When high temperatures are required, and heat transfer is mainly by radiation, as for example in melting and reheating furnaces used in the steel industry, flame luminosity and emissivity associated with liquid fuels have obvious advantages.

Factors affecting Choice

The true cost of a fuel for any particular process is not, however, its basic cost/therm but the overall cost per unit of production.

Factors affecting Overall Cost

1. Basic cost of fuel delivered to user's premises.
2. Overheads involved in storing fuel (if necessary) and delivering it in right condition at the appliance combustion chamber. (See Section 4.)

Capital Costs: These include
  Storage tanks
  Heaters and pumps
  Gas mains. Electric cables. Meter
  Meter house or sub-station
  Mechanical stokers
  Fans and controls

Operating Costs:
  Labour
  Maintenance
  Preheating and pumping
  Value of fuel storage site
  Ash disposal

3. Average thermal efficiency of plant expressed in fuel used per unit quantity of production over a period.
4. Incidental factors including: —
  Environmental conditions
  Quality of final product
  Amount of handling required
  Automation of plant — integration into flow line production
  Maintenance
  Service available from fuel supplier.

Often it is advisable to have an economic and technical appraisal made by an expert and impartial authority of all the factors involved before deciding on the fuel to be used.

FACTORS INFLUENCING THE EFFICIENCY AND PERFORMANCE OF FURNACES

Design Considerations

The aim of the furnace designer should be to obtain optimum operating efficiency from the furnace and therefore care must be taken to match the shape and dimensions as closely as possible to the requirement of the material to be processed.

Although no fixed rule can apply, important factors that need careful consideration include: —

1. Type, size and shape of material to be processed.
2. Temperature requirements and heating cycle.
3. Size and configuration of furnace chamber.
4. Furnace throughput or melting rate.

5. Batch or continuous operation.

6. Furnace atmosphere requirements.

It is of primary importance that the designed capacity of the furnace is closely related to the required output since one of the main causes of inefficiency lies in incompletely filled furnaces. For example, a furnace designed for an output of 5 tons per hour will have the same structural heat losses if only 1 ton per hour is being produced.

The shape of the furnace chamber is also of vital importance and this should be so designed to permit the maximum amount of radiation to be transmitted to the work by the hot furnace surfaces and to encourage recirculation of hot gases.

The number, size and disposition of the burners or burner ports requires careful attention if rapid attainment of the required temperature, and uniformity of heating without flame impingement is to be achieved. The size and number of flue offtakes is equally important and they should be located relative to burner positions to give a satisfactory hot gas flow pattern and to ensure optimum travel of the gases before their final discharge to atmosphere. Ideally all parts of the work chamber should be filled with moving products of combustion.

Special attention should be given to the design of the furnace door, for a badly designed door is inevitably a source of considerable heat loss. For example, a badly-fitting door on a furnace would permit considerable inleakage of cold air into the furnace chamber if operating under suction conditions. Not only would this increase fuel consumption but cold spots would develop in the vicinity of the door and this could have harmful effects on the material being heated.

HEAT STORAGE CAPACITY AND STRUCTURE HEAT LOSSES

The choice and use of the most suitable type of refractory and insulating material in the construction of the furnace has a vital influence on fuel economy and furnace performance.

Where furnaces operate intermittently it is necessary, if high thermal efficiency is to be achieved, to use materials and furnace wall thicknesses properly related to the hours of use, required temperature, operating cycle and hearth loading rates. Too heavy a construction and over-insulation increases heat storage capacity and is wasteful of fuel. The amount of insulation to be used is a simple matter of economics.

Two types of material are now widely used in the construction of small furnaces: —

1. *Hot Face Refractory Insulation*

   This material has a much lower bulk density than firebrick and, as a result, much lower heat capacity and a reduced thermal conductivity figure, i.e. 0.335 K Cal/sq metre/h/°C/metre (2.7 Btu/sq ft/h/°F/ inch) compared with a figure of 0.99/1.2 K Cal/sq metre/h/°C/metre for firebrick (8/10 Btu/sq ft/h/°F/ inch) considerable fuel savings can be made.

2. *Ceramic Fibre*

   Ceramic fibre is made from high purity alumina-silica china clay.

   It is an ideal hot face lining for small furnaces operating at temperatures up to 1260°C.

   In its blanket form it is easy and quick to install and due to a very low thermal conductivity figure which is 35% better than insulating firebrick and as it is one-sixth of the weight, heat storage and losses are substantially reduced. Additional advantage is that heating-up and cooling times can be reduced by 30%. Thickness of ceramic fibre normally used ranges between 25 mm (1″) to 75 mm (3″) with a backing of suitable insulating material. A careful appraisal is needed to ensure the most economic thickness.

Diagram 1 shows heat storage capacities and heat losses of a furnace wall made up of composite materials of varying thicknesses and assumes the furnace to be operating intermittently at 1000°C for 12 hours a day, 5 days per week. The calculations are based on a wall area of 3.716 sq metres (40˙ sq ft).

The saving in heat input as indicated in the Diagram is significant but it should also be borne in mind that as a production unit, the furnace with the lower thermal capacity will reach temperature much quicker than one with a heavier construction.

When furnacce operation is continuous the thermal capacity of the brickwork is not so important and the main objective should be to minimise heat losses from the structure by an adequate insulating backing.

292

Diagram 1

Thermal capacity and heat losses through various wall constructions from a furnace at 1000°C and having 3.716 m² (40 sq ft) of internal surface area and operating over a 5-day weekly period intermittently.

Wall Construction	Intermittent Operation 12 h/day 5 days/week			
	9″ (229 mm) Firebrick	9″ (229 mm) Hot Face	4½″ (114 mm) Hot Face +4½″ (114 mm) Diatomaceous Insulation	2½″ (63 mm) Ceramic Fibre Lining +4″ (102 mm) Mineral Wool
Heat Loss through Brickwork MJ/week (therms/week)	3217 (30.5)	1192 (11.3)	722 (6.84)	443 (4.2)
Heat Storage Loss after shut-down MJ/week (therms/week)	5275 (50.0)	1720 (16.3)	1783 (16.9)	385 (3.65)
Total Heat Loss MJ/week (therms/week)	8493 (80.5)	2912 (27.6)	2505 (23.74)	828 (7.85)
% Reduction in Heat Losses by using insulation	—	65.5	70.5	90.25

COMBUSTION EQUIPMENT

The selection of a suitable burner for a particular application is of paramount importance if optimum operating efficiency and performance is to be achieved.

The burner should be an integral part of the furnace design and its characteristics should closely conform to the required heat flow and distribution.

Important factors that need careful consideration include: —

Capacity and number of burners required
Type of flame needed
Burner turn-down ratio
Furnace atmosphere required, i.e. oxydising,
neutral or reducing
Ease of ignition and maintenance
Safety aspects
Noise emission

The type of burners most commonly used for small furnace applications are briefly described below: —

Gas Burners

The trend in gas burners has been away from the simple diffusion flame and atmospheric types towards more sophisticated and accurately controlled premix and nozzle mixing burners incorporating air blast, linked valve or mechanically premixed systems.

Premix Burners using Pressure Air

In this type of burner, air and gas is mixed in an injector upstream of burner ports.

Air is provided by a low pressure fan and gas is entrained into the air supply from a zero governed supply.

These burners possess an inherent self proportioning action and this enables a fairly constant air/gas ratio to be maintained over the operating range by adjustment of a single valve in the air line.

Nozzle Mixing

Air and gas are kept separate in this type of burner and do not mix till they meet at the burner port.

Here again air is supplied from a low pressure fan and the burner has high turn down characteristics.

294

The present trend is towards the use of packaged and high velocity burners.

A packaged burner includes all required ancillaries as an integral system and is generally manufactured in its entirety at the factory.

The refractory tunnel used with a high velocity burner is of sufficient size to allow combustion to be completed within the tunnel. The products of combustion are discharged at high velocity from the exit.

High velocity firing promotes turbulence and high recirculation rates in the furnace chamber. In consequence heat transfer rate is greatly improved and better temperature uniformity is obtained. The use of high velocity burners generally enables the number of burners to be reduced.

A burner that has recently been designed and developed is the recuperative burner. In a single unit this incorporates a high velocity burner, flue and recuperator.

Combustion air is preheated in the integral recuperator by waste gas, and air temperatures of up to 600°C are possible.

A number of field trials have been undertaken with recuperative burners on conventional furnaces and results have shown fuel savings in the order of 30-50%; there appears to be tremendous potential for the future application of this type of burner.

Oil Burners

In burning liquid fuels it is important that the oil be delivered to the burner at correct viscosity for atomisation, i.e. 100 secs Redwood No. 1.

Oil and atomising air pressures should always be in accordance with burner manufacturers' recommendations.

Low Pressure Air Burners

This type of burner is supplied with combustion and atomising air from a low pressure fan. In its simplest form it is equipped with separate controls on both oil and air supplies.

An improved type is the self proportioning burner and control can be obtained by movement of a single lever which simultaneously regulates and proportions the oil and air supply to maintain satisfactory air/fuel ratios over the turn-down range.

Medium Pressure Air Burners

These are supplied with air from a rotary-type compressor at pressures between 0.138 Bar and 1.035 Bar (2 and 5 p.s.i.). Only a small proportion of the combustion air is supplied at this pressure for atomising the oil (usually about 5%) the remainder being induced or supplied by a fan at a pressure of a few inches w.g.

With this arrangement a high turn-down ratio can be achieved without loss of combustion efficiency. A further advantage is that the bulk of the combustion air can be pre-heated to as high a temperature as possible.

As with gas burners, there is an increasing trend with oil burners towards more sophisticated types including packaged and high intensity burners generally on the lines previously discussed under gas burners.

The installation of dual fuel burners may be a wise investment in view of: —

(a) recent history of uncertainty of fuel supplies
(b) to enable advantage to be taken of interruptible supplies if available and
(c) to take advantage of fuel price differentials.

OPERATION AND CONTROL

Control of Heat Input

The maximum heat input should be carefully related to the temperature and heating rate required and design limitations of the furnace particularly with regard to combustion space volume and capacity of exhaust system.

Excessive fuel input can result in incomplete combustion with high flue gas loss and wastage of fuel. In addition, over-firing can lead to excessive wear of refractories with possible damage to stock.

Both the fuel and air supplied to the furnace should be capable of easy regulation and in the case of oversized burners the air and fuel supplies should be restricted to required capacity.

Combustion Control

To achieve optimum combustion efficiency and satisfactory air/fuel ratio it is essential to burn the fuel with the minimum amount of excess air.

296

Accurate maintenance of the correct air to fuel ratio is probably the most important single factor in contributing to good operating efficiency. As far as is practicable the optimum CO_2 should be maintained in the furnace gases at all rate of fuel input.

If air in excess of the normal requirements is used to burn the fuel, efficiency will be lowered and wastage occurs, because this excess will be heated and carried away as additional sensible heat in the waste gases. This effect of excess air on amount of heat carried away together with corresponding CO_2 percentages is shown in Diagrams 2 and 3. In addition to increasing waste gas loss, excess air reduces flame temperature and adversely affects heat transfer and furnace atmosphere. Consequently, the amount of air used should always be the smallest quantity that will enable the fuel to be properly burned in the prevailing conditions.

Diagram 4 shows the effect of excess air on flame temperature.

It is essential that correct combustion conditions be maintained throughout the entire turn-down range of the combustion system to ensure minimum fuel wastage.

Excess air can also have harmful effects on the quality of product, i.e., high free oxygen levels can lead to excessive metal loss and surface damage to stock.

Insufficient combustion air also results in fuel waste and the reducing furnace atmosphere produced, i.e., oxygen free, can be very deleterious on certain metals, e.g., hydrogen pick-up in aluminium melting/holding, or the surface defects on titanium.

Regular check of CO_2 content in waste gases should be made to ensure that optimum combustion conditions are being maintained and it should never be assumed that once a furnace has been set up correctly, no further attention is needed.

When checking flue gas analysis it is important that the sampling position chosen is one that will give a truly representative figure of combustion conditions as a whole.

Most combustion systems now incorporate air/fuel ratio control to maintain combustion quality. These include linked valves, combined adjustable port proportional valves, pressure and volumetric governing techniques.

297

APPROX. HEAT LOSS TO WASTE GAS FROM DRY
PRODUCTS OF COMBUSTION AS % OF NET C.V.

FUEL OIL

20% Excess Air CO_2 — 13% approx.

Waste Gas Temp.	Loss as % of Heat in Fuel
100°C	4.0
200°C	9.0
400°C	19.0
600°C	29.0
800°C	39.0
1000°C	49.0
1200°C	59.0

with Waste Gas at 1000°C

Waste Gas CO_2	% Excess Air	Loss as % of Heat in Fuel
13.0%	20	49.0
11.0%	40	56.0
9.7%	60	64.0
8.5%	80	71.5
7.6%	100	79.0
6.9%	120	86.5

APPROX. HEAT LOSS TO WASTE GAS FROM DRY PRODUCTS OF COMBUSTION AS % OF NET C.V.

NATURAL GAS

10% Excess Air CO_2 — 10.7%

Waste Gas Temperature °C	Loss as % Heat in Fuel
100	3.5
200	8.0
400	16.5
600	23.5
800	31.5
1000	41.0
1200	50.0

With Waste Gas at 1000°C

Waste Gas CO_2	% Excess Air	Loss as % of Heat in Fuel
10.7	10	41.0
9.8	20	45.0
8.3	40	53.0
7.2	60	60.5
6.3	80	68.0
5.7	100	75.5

APPROX. EFFECT OF EXCESS AIR
ON FLAME TEMPERATURE

% Excess Air	Flame Temperature °C (Fuel Oil)
Nil	2080
20	1870
50	1630
80	1450

Temperature Control

Close control of the time cycles and temperature conditions is essential in modern furnaces.

Manual control by visual observation can vary widely according to operator's interpretation and this inevitably leads to increased fuel consumption with possible product damage by over or under-heating and excessive oxidation.

The provision of automatic control is therefore most important and here the equipment can vary widely from the simple two position indicating controller to the more elaborate and sophisticated type of instrument which provides modulating control closely related to furnace heat demand, thus avoiding large and rapid changes in heat input.

Where heating or cooling rates are important an automatic temperature/programme controller can be used to control the furnace to a predetermined time/temperature schedule. Thus, once the load has been charged into the furnace, heating up, soaking and cooling is automatically controlled. Finally, automatic control o ftemperature will always give substantial benefits.

The degree of sophistication introduced into the control system will depend to a large extent on the value of the product being processed and furnace running costs.

Furnace Pressure Control

For good practice, pressure in the furnace working chamber must be slightly in excess of atmospheric, i.e. about +0.05 m bar (0.02 ins w.g.) at all rates of heating.

Negative pressure conditions permit the inleakage of cold air through doors and other openings. This cools the furnace, creates cold spots, increases waste gas losses and drastically affects furnace atmosphere conditions, again often resulting in excessive oxidation of the charge.

300

Excessive pressure on the other hand may retard combustion development and increase losses by flame and hot gas emission through furnace openings.

Slightly positive pressure conditions can be maintained by careful adjustment of flue dampers. The function of a damper is to regulate the draught requirements for the various rates of heat input.

Correct regulation of the damper is of extreme importance if efficient fuel utilisation is to be achieved, but it is often the most neglected part of the furnace and its malfunction is frequently the cause of considerable fuel wastage.

In practice, manual operation of the damper to meet the changing requirements of the heating operation is difficult. This is particularly so with oil and gas fired furnaces which operate under high/low conditions.

The provision of an automatic furnace pressure controller is nearly always a worthwhile investment for not only can fuel economies of around 10% often be achieved by maintaining correct pressure conditions but in addition temperature uniformity and furnace atmosphere conditions will be better resulting in quicker heating up times and improved product quality.

INSTRUMENTATION

Adequate instrumentation is an essential aid for the maintenance of optimum furnace efficiency and performance and even more important in the majority of cases, to ensure process requirements are fully met.

Instruments are used

Control temperature/time cycle
Ensure correct air/fuel ratio is maintained at all rates of heat input
Obtain required furnace atmosphere
Ensure fuel is used economically
Measure and/or record fuel input rate.

One of the best checks on the working of a furnace is to determine and record at regular intervals the amount of fuel used/unit of output. This necessitates the installation of flow meters, but this is usually a wise investment. Comparison of any given week's performance with that of preceding weeks is a valuable check of furnace efficiency. Any falling off in performance should be investigated and the aim should be to continuously improve on past records.

Full instrumentation is initially expensive but experience has shown that a substantial return is obtained in fuel economy, consistent production with minimum rejects, and in reduced furnace maintenance costs. Instrumentation can of course be further developed to incorporate automatic controllers which reduce the work of the operator and can make adjustments much more frequently than any operator could hope to do.

FURNACE UTILISATION

Productive Use of Furnace Heat

When the initial design of a furnace is considered discussions should take place with all parties involved in the process, of which the furnace is a part, the aim being to obtain the correct furnace for the jobs in hand. One factor which must be considered is whether or not the work can be processed in a continuous furnace or must a batch type be used.

If stock can be fed continuously into one end of a furnace and discharged at the other then the overall efficiency will increase due to load recuperation from the waste gas products.

If it is not possible to use a continuous furnace, then careful planning of the loads for batch type furnaces is essential. A furnace should be re-charged as soon as possible (within the metallurgical limitation) to enable any residual furnace heat to be used.

Furnace Loading

One of the most vital factors affecting efficiency is loading. There is a particular loading at which the furnace will operate at maximum thermal efficiency. If the furnace is underloaded a smaller fraction of the available heat in the working chamber will be taken up by the load and therefore efficiency will be low.

The best method of loading is generally obtained by trial — noting the weight of material put in at each charge, the time it takes to reach temperature and the amount of fuel used. Every endeavour should be made to load a furnace at the rate associated with optimum efficiency although it must be realised that limitations to achieving this are sometimes imposed by work availability or other factors beyond control.

Diagram 5

EFFECT OF HEARTH LOADING ON THERMAL BALANCE AND SPECIFIC FUEL CONSUMPTION

Slot Type Reheating Furnace for Drop Forging

Exit gas conditions — 1320°C and 10% excess air
Total heat to structure — 264 MJ/h (2.5 Therms/h)
Effective Hearth Area — 1.277 m² (13.75 sq ft)
Heat in Gases leaving furnace — 60%
Heat available in furnace — 40%

		Hearth Loading (kg/m²/h) (lb/sq ft/h)			
		98 (20)	147 (30)	244 (50)	366 (75)
Furnace Output	kg/h (ton/h)	125 (.123)	187 (.184)	312 (.307)	468 (.460)
Heat in Steel @ 1250°C = Hc	MJ/h (therms/h)	108 (1.025)	162 (1.54)	270 (2.56)	406 (3.85)
Heat to Furnace Structure = Hs	MJ/h (therms/h)	264 (2.5)	264 (2.5)	264 (2.5)	264 (2.5)
Total Heat Required $\dfrac{Hc+Hs}{0.4}$ = HI	MJ/h (therms/h)	928 (8.8)	1066 (10.1)	1329 (12.6)	1688 (16.0)
Ratio: Heat to Steel $\dfrac{Hc}{Hs}$ Heat to Structure		0.41	0.62	1.02	1.54
Efficiency of Steel Heating $\dfrac{Hc}{HI}$	%	11.6	15.2	20.3	24.0
Fuel Consumption	MJ/tonne steel (therms/ton steel)	7543 (71.5)	5803 (55.0)	4347 (41.2)	3671 (34.8)
Fuel and Financial Saving	%	—	23.0	42.0	51.0

Diagram 6a

EFFECT OF LOADING ON CONTINUOUS FURNACE PERFORMANCE

Production Rate	Tonnes/h Ton/h	10.0 (9.8)	5.0 (4.9)	3.0 (3.0)	1.0 (1.0)
Heat to Stock	MJ Therms	7385.0 (70)	3692.0 (35)	2215.0 (21)	738.0 (7)
Waste gas Loss	MJ Therms	10550.0 (100)	7385.0 (70)	6330.0 (60)	5412.0 (51.3)
Structure Loss, etc.	MJ Therms	7385.0 (70)	7385.0 (70)	7385.0 (70)	7385.0 (70)
Total Heat Input	MJ Therms	25320.0 (240)	18462.0 (175)	15930.0 (151)	13535.0 (128.3)
	MJ/Tonne	2532.0	3692.4	5310.0	13535.0
	Therms/t	24.5	35.7	50.3	128.3

Diagram 7 shows test results on another reheating furnace and again illustrates the high "no load" heat input needed, and the advantage of operating at the highest economic furnace loading.

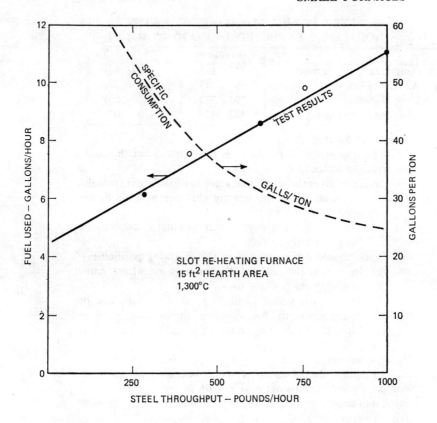

Diagram 6B

The "no-load" consumption to maintain a simple batch type furnace at 1300°C when empty is about 70% of that required to operate at optimum loading rate.

Diagrams 5, 6a and 6b indicate the effect increased loading rates have on the thermal efficiency and performance of a typical slot type reheating furnace for drop forging and of a continuous reheating furnace. Diagram 7 shows typical hearth loading rates normally expected in good practice.

305

TYPICAL HEARTH LOADING RATES
NORMALLY EXPECTED IN GOOD PRACTICE

	$kg/m^2/h$	(lb/sq ft/h)
Heat Treatment Furnace	147 – 195	(30 – 40)
Annealing Furnace	195 – 293	(40 – 60)
Drop Stamping and Forging	293 – 390	(60 – 80)
Continuous Reheating	342 – 489	(70 – 100)

Placing of Stock

The disposition of the load on the furnace hearth should be arranged so that: —

(a) It receives the maximum amount of radiation from the hot surfaces of the heating chamber and the flames produced.

(b) The hot gases are efficiently circulated around the heat receiving surfaces.

Stock should not be placed in the following positions: —

(a) In the direct path of the burners or where flame impingement is likely to occur.

(b) In an area which is likely to cause a blockage or restriction of the flue system of the furnace.

(c) Close to any door openings where cold spots are likely to develop.

Load Residence Time

In the interests of economy and work quality the materials comprising the load should only remain in the furnace for the minimum time to obtain the required physical and metallurgical requirements.

When the materials attain these properties they should be removed from the furnace to avoid damage and fuel wastage.

The higher the working temperature the higher the loss per unit time.

The effect on the materials by excessive residence time will be an increase in surface defects due to oxidation. The rate of such oxidation is dependent upon time, temperature, as well as free oxygen content.

This possible increase in surface defects can lead to rejection of the product or expensive secondary surface techniques such as turning, grinding, polishing, etc.

It is therefore essential that liaison between the furnace operator, production and planning personnel be maintained.

WASTE HEAT RECOVERY

The major loss of heat from fuel fired furnaces is that which is carried away as sensible heat in the waste gases. With furnace temperatures varying from say 800° to 1400°C this loss ranges from 40% to 70% of the heat input.

By recovering and utilising some of this waste heat, overall efficiency can be greatly improved and substantial fuel savings achieved.

Waste heat can be recovered by: —
1. Preheating combustion air through recuperators or regenerators.
2. Preheating the load before it enters the working chamber by outgoing waste gases — load recuperation.
3. External use, for steam raising, space heating, water heating or other low temperature processes.

Combustion Air Preheat

The most effective method of recovering some of the waste heat leaving the furnace is to preheat the combustion air. This method enables the waste heat to be used at all times the furnace is in operation.

There are two different types of heat exchangers: —
(a) Recuperators
(b) Regenerators

Regenerators are only normally used on large high temperature furnaces and are therefore not further considered.

Recuperators

A recuperator is a heat exchanger between the waste gases and the gas to be preheated. It usually consists of a system of ducts or tubes some of which carry the air for combustion to be preheated, the others containing the source of waste heat. Recuperators may be of either metallic or refractory construction, according to the temperature and the corrosive properties of the waste gases involved.

A type of metallic recuperator commonly used with small high temperature furnaces consists of two concentric tubes. Combustion air flows between the annulus between the tubes either in parallel or counter flow and heat transfer from the waste gas is mainly by radiation.

Refractory recuperators generally have limited use because of difficulties of maintaining tight air/waste gas seals and also due to their susceptibility to thermal shock.

It is important when installing a recuperator in the flue system to ensure that draught loss is minimised otherwise furnace operation and performance may suffer.

Diagram 8 shows possible savings that can be achieved by preheating combustion air.

POTENTIAL SAVINGS USING AIR PREHEAT

Gaseous Fuels   10% Excess Air

Air Preheat	Savings
100°C	4.0%
200°C	8.5%
300°C	13.5%
400°C	17.0%
500°C	20.2%

Fuel Oil   20% Excess Air

Air Preheat	Savings
100°C	5.5%
200°C	11.2%
300°C	16.2%
400°C	20.5%
500°C	24.0%

Factors Affecting the Choice of
Heat Exchanger

1. Temperature of the waste gas.
2. Condition of the waste gas, i.e. contamination from solid or gaseous particles.
3. Air preheat temperature required.
4. Type of burner used on furnace and its limitation for air preheat.
5. Control required, i.e. high air preheats will need compensatory systems to maintain the correct air/fuel ratios.
6. Possible secondary use of the waste heat after leaving the primary heat exchanger.
7. Capital cost and recovery period.
8. Possible need to use extra fans and control equipment due to pressure losses.

Load Recuperation

Load recuperation is another effective means of heat recovery and is more commonly used on continuous furnaces where the outgoing hot gases preheat the incoming load. Examples where load recuperation is practised include pusher type, inclined hearth and roller hearth furnaces.

Compared with batch type furnaces thermal efficiencies of continuous furnaces are generally considerably higher.

Load recuperation can also be achieved by utilising a preheating chamber. In this case the cold load is preheated by the waste gas before it is inserted into the main furnace.

External Use of Waste Heat

If waste heat is not to be used for preheating combustion air or load recuperation, its utilisation for space heating, water heating and low temperature process such as drying, warrants careful consideration; providing scope exists, appreciable economies can be gained.

The merits of each individual case for waste heat recovery, however, require careful investigation and the possibilities of furnace operation and performance becoming subordinate to waste heat utilisation should be borne in mind.

Factors Influencing Installation of
Recovery Equipment

Before installation of waste heat recovery equipment an economic appraisal must be made.

The factors to be considered are: —

(a) The quantity of recoverable heat in the waste.
(b) The most advantageous way of using the heat.
(c) Does the availability match up to requirements?
(d) Capital and running costs of plant required for heat extraction.

An economic solution can be found only when all the facts are known and one process is related to another.

CONCLUSION

From the foregoing notes it will be appreciated that the achievement of optimum furnace efficiency and performance is dependent on many factors.

The basic principles of economic furnace performance are good design and the incorporation to a maximum extent of insulating materials, maximum combustion and operating efficiency, instrumentation, regular maintenance, recovery of waste heat whenever possible, and optimum loading. It has been emphasised that one of the major causes of inefficiency is the failure to ensure maximum utilisation of the furnace capacity.

By careful attention to the items mentioned, furnace efficiencies can often be greatly increased and fuel savings achieved. The extent of these savings vary considerably, but experience shows that on average the figure lies between 10 and 30%.

Diagram 9 shows a summary of recommendations made after tests on 250 forge furnaces. It will be seen that certain recommendations occur frequently, and that the potential savings on all these furnaces is 21% of the combined fuel usage.

Another way of expressing Diagram 9 is shown in Diagram 10 where a typical furnace is shown "as found" and "as improved" with suitable notes to refresh the reader's memory on the possible improvements to furnaces that have been covered in this section.

Draught problems in furnaces can also make it difficult to improve efficiency and Diagram 11 shows improvements made to an existing pusher type furnace for billet heating in which the addition of an adjustable curtain wall made it possible to operate under very slight pressure in the high temperature zone, eliminates air infiltration and also encourage better "scrubbing of the billets by the gases leaving the furnace.

A recuperator was also installed to preheat the combustion air. The efficiency of one furnace that was converted in this way was increased from 30% to 42% which represented a reduction in heating costs of some 29% also the thickness of the surface layer of metal effected by decarburization was reduction in heating costs of some 29%; also the thickness of no work has been rejected since the modification.

310

1 No burner quarl. Unsuitable burner gives poor combustion efficiency.

2 Poor design gives uneven heat distribution.

3 Excessive height relative to furnace loading.

4 Poor loading gives limited use of available heat.

5 Construction in heavy refractory material with high heat losses.

6 Insufficient control of furnace temperatures.

7 No flue or recuperator. Hot gases at 1,300°C exhaust through charging slot.

AS FOUND

1 Specially designed burner quarl. Correct burner using preheated combustion air.

2 Improved circulation of hot gases due to redesigned furnace chamber.

3 Volume of furnace reduced by 40%. Area of hearth reduced by 25%.

4 Maximum use of slot width and improved loading and soaking.

5 Thermal insulation using light-weight refractories reduce heat losses.

6 Automatic temperature control installed.

7 Installation of recuperator to preheat combustion air.

AS IMPROVED

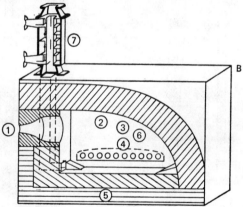

Diagram 10

311

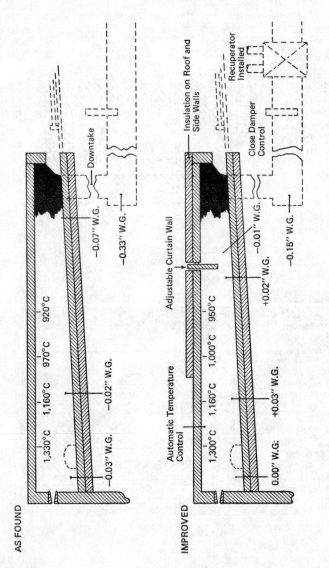

AS FOUND

1,330°C 1,160°C 970°C 920°C

−0.03″ W.G. −0.02″ W.G. −0.07″ W.G. −0.33″ W.G.

Downtake

IMPROVED

Automatic Temperature Control

1,300°C 1,160°C 1,000°C 950°C

0.00″ W.G. +0.03″ W.G. +0.02″ W.G. −0.01″ W.G. −0.15″ W.G.

Adjustable Curtain Wall

Insulation on Roof and Side Walls

Close Damper Control

Recuperator Installed

Diagram 11

312

FREQUENCY OF RECOMMENDATIONS		
250 Forge Furnaces		
Replace burners by smaller self-proportioning types	:	125
Improve oil preheat	:	52
Use waste heat to pre-heat combustion air	:	150
Improve combusion control	:	132
Reduce excess furnace temp.	:	139
Operational adjustments	:	117
Modify furnace design	:	196
Incorporate insulating brick	:	96
*Improve planning and loading	:	143
Improve instrumentation	:	12
Install ring main system	:	19
Improve repair and maintenance	:	27
*Saving excluding * estimated*	:	21%

Further examples of energy conservation

1. A gas fired bogie hearth heat treatment furnace for hardening and tempering of large steel castings at 900°C and 620°C respectively was tested and found to have a very low thermal efficiency of 7%. It was found that by lowering the hearth by 6″ (15cm) two castings could be accommodated instead of one; combustion control was installed; the stack damper was fitted with an automatic actuator; insulation was added to the furnace roof; the existing recuperator was extended to double the heat exchanger surface, and other minor alterations were made. The reduction in gas usage per ton of castings was 61%, or £58,000 per year with gas at 18 pence per therm.

2. An aluminium melting and alloying furnace was used on a 24 hour cycle to melt metal and then hold it in molten state until required by the foundry. Oil usage as found was 630 gallons per cycle. Solely by improving combustion control and damper operation this was reduced to 522 gallons per cycle, a 17% saving. However it was clear that further savings were possible. The melting period was extended in time by 35% and only two of the four oil burners then were

313

necessary. These could then operate near the upper end of their design, with improved combustion, higher flame temperature and resultant increased heat transfer efficiency. Air leakage was almost eliminated by operating under slight pressure in the furnace. This reduced oil consumption per cycle to 475 gallons, increasing overall savings to 24.6%. While these savings had been obtained without appreciable capital expenditure, it was clear from the tests that had been made that further potential savings existed. Instruments to record CO_2 and draught, and a recuperator to preheat the combustion air, were installed, and the cycle consumption is now 400 gallons, a reduction of 36.5% of the initial fuel costs.

3. A new enclosure was designed for some lead melting pots, using well known principles for combustion-chamber design, and providing adequate insulation. The result has been an increase of 20 per cent in production and 30 per cent saving in fuel. As in many similar processes, the metal melting rate set the pace of production throughout the works; thus the importance of this improvement to the firm was considerable.

4. When investigating conditions on some small gas fired melting pots, a visual examination showed nothing untoward, but analyses of the flue gases give figures of 2 per cent CO_2. By the simple expedient of sealing the space beneath the pot and restricting the exhaust gas flow, excess air was reduced from 475 to 80 per cent; gas consumption immediately fell from 456 ft^3/h to 210 ft^3/h, a reduction of 54 per cent. Melting times were reduced as a result of improved performance.

5. A large salt bath was taking 70 hours to heat from cold. Again, analysis of the flue gases was undertaken. This showed that, due to a restriction at the air inlet, combustion was incomplete, and 4 per cent CO was present in the flue gases. A small modification, quickly carried out, resulted in heating time being reduced by 50 per cent and gas consumption by 645 therms/week.

6. An investigation of a furnace for annealing non-ferrous metals serves as a good illustration of the improvements made in this field. A gas-fired recuper-

ative furnace was in constant use, six days a week, for annealing coiled strip brass at 650° and 700°C.

By carrying out the recommendations listed below, gas consumption per ton of metal heated was reduced by 23 per cent. In addition, the time required to heat a batch of metal was reduced from 180 to 155 min; output was thus increased by 16 per cent, with an improvement in quality.

Recommendations: (a) Refit furnace door; (b) Eliminate air inleakage; (c) Overhaul temperature control system; (d) Install gas pressure governor; (e) Adjuit combustion-air controls; (f) Reduce size of burner jets.

In many such instances the value of increased output may be much greater than that of the fuel saved.

Final Conclusions

The ten sections jointly written by many of the senior staff of NIFES, have attempted to give practical guidance on conserving energy. This has been the main object, and those who have expected a great deal of descriptive or theoretical material may feel disappointed.

Although NIFES have academic resources it is not an "academic" institution and has obtained its experience by inspection and tests carried out for thousands of clients. Some may complain that there are gaps in coverage, and it has certainly been impossible to discuss results on every type of energy-using plant, but the general principles of energy conservation still apply.

Since the Introduction to this Handbook was written in early 1974 fuel prices have continued to rise, and at the beginning of 1975 the price of fuel oils to many customers had risen to around five times the 1972 figure. The price of coal, gas and electricity has not risen so steeply, but Government pressure on these industries to eliminate losses and subsidies will cause these prices to climb until the present disparity with oil is much reduced. Although "North Sea" oil will reduce the balance-of-payments problem in the late 1970's

315

and 1980's, the massive investment in exploration and production facilities preludes any hope of a price reduction.

The advice of this Handbook can finally be summed up in the last two Tables. Table A shows the essential steps of any energy conservation programme, whether on a single item of plant or a whole factory, and emphasises that "selling" recommendations to production and financial management is just as important as the investigation and decision-making. Many a good idea has failed to be implemented because it was not put forward in terms that financial management could understand. Even after successful implementation, follow-up work and setting of targets is essential to prevent gradual erosion of the savings and benefits.

Table B is to remind the reader that there are categories of saving and in difficult times, when capital is scarce, managements may be reluctant to invest in new plant — it may quite literally be a case of "economy at any cost"!

Table A
EFFICIENT USE OF ENERGY

Essential steps: —

1. IDENTIFICATION — of areas of waste.
2. INVESTIGATION — including measurement, to establish facts.
3. QUANTIFICATION — of losses and value.
4. DECISIONS — on possible courses of action and choice of best.
5. PRESENTATION — of recommendations, of reasons for making them, and financial advantages.
6. IMPLEMENTATION — ensure plan is followed.
7. FOLLOW-UP — check that estimated savings have been obtained. Did alterations have any unforeseen effects?
8. SET TARGETS — to ensure new performance is maintained.
9. RE-EXAMINE (at intervals) — Can further improvements be made? Do alterations in production make any modifications necessary?

EFFICIENT USE OF ENERGY
Table B

Categories of Saving: —
1. IMPROVED OPERATION AND MAINTENANCE
   (Revenue Investment)
2. MODIFICATIONS AND ADDITIONS TO EXISTING
   PLANT   (Small Capital Investment)
3. INSTALLING NEW PLANT
   (Considerable Capital Outlay)
4. SUBSTITUTION OF EXISTING PROCESS
   BY NEW METHODS, ETC.
   (Long-term or Major Capital Outlay).

In conclusion it may be well to sound a note of warning. The reduction in manufacturing output of the industrial nations which is taking place at the moment will cause a flattening off in world energy demand, which could effect a temporary lowering of prices. Energy prices however, will never descend to the level that we have known in the 1960-70 period. The era of cheap energy is over and until such times as the long term alternatives to fossil fuels such as Nuclear Energy, Solar power, etc., are with us, Energy Conservation is the only alternative, and it pays the Industrialist and the Nation.